The New Zealand Romney Marsh Sheep Flock Book
Volume 3

by New Zealand Romney Marsh Sheep Breeders Association

with an introduction by Jackson Chambers

Self Reliance Books

Get more historic titles on animal and stock breeding, gardening and old fashioned skills by visiting us at:

http://selfreliancebooks.blogspot.com/

Introduction

I am pleased to present yet another practical title on breeding and raising livestock.

The work is in the Public Domain and is re-printed here in accordance with Federal Laws.

As with all reprinted books of this age that are intended to perfectly reproduce the original edition, considerable pains and effort had to be undertaken to correct fading and sometimes outright damage to existing proofs of this title. At times, this task is quite monumental, requiring an almost total "rebuilding" of some pages from digital proofs of multiple copies. Despite this, imperfections still sometimes exist in the final proof and may detract from the visual appearance of the text.

I hope you enjoy reading this book as much as I enjoyed making it available to readers again.

Jackson Chambers

CONTENTS.

The New Zealand Romney Marsh Sheep Breeders' Association.

OFFICERS AND COUNCIL.

OFFICERS:

President—Mr. G. C. WHEELER, Burford, Stanway.

Vice-President—Mr. H. V. FULTON, Crawford Street, Dunedin.

Hon. Treasurer—Mr. W. F. JACOB, Kiwitea, Feilding.

COUNCIL:

Mr. T. P. ALLEN, Walmer Farm, Waiwetu, Lower Hutt.

Mr. J. O. BATCHELAR, Palmerston North.

Mr. R. N. BEALEY, Haldon, Hororata, Canterbury.

Mr. D. P. BUCHANAN, Mayfield, Cunninghams.

Mr. R. GRAY, Fairburn, Masterton.

Mr. J. W. HARDING, Mount Vernon, Waipukurau.

Mr. A. E. HARDING, Mangawhare, Auckland.

Mr. J. HOLMS, Waimahaka, Invercargill.

Mr. F. HUTCHINSON, JUNR., Rissington, Hawke's Bay.

Mr. J. KEBBELL, Te Rauawa, Ohau, Manawatu.

Mr. A. MATTHEWS, Waiorongomai, Featherston.

Mr. E. SHORT, Parorangi, Feilding.

Bankers—BANK OF NEW ZEALAND.

Secretary—E. J. WACKRILL, P.O. Box 40, Feilding.

Registered Office of the Association—Fergusson St., Feilding, N.Z.

MEMBERS.

ALLEN, JOHN, The Cliffs, Waingaro, Waikato.
AKERS, W., Riversdale, Palmerston North.
ALLEN, E. W., Dealwood, Clareville, Wairarapa.
ALLEN, W. B., Stoneleigh, Clareville, Wairarapa.
ALLEN, G. E., Stoneleigh, Clareville, Wairarapa.
ALLEN, T. P., Walmer Farm, Waiwetu, Lower Hutt.
ALLEN, J. C., Annandale, Piako, Auckland.
ANSON, DR., Te Kumu, Mangaonoho.
AITKEN, W. G., Ashhurst.
BATCHELAR, JOSIAH, Brookleigh, Linton.
BUICK, D., Cloverlea, Palmerston North.
BUCHANAN, D. P., Mayfield, Cunninghams.
BULL, G. W., Waikiekie, Auckland.
BEALEY, R. N., Haldon, Hororata, Canterbury.
BEYERS, H., The Brow, Waipawa.
BIDWILL, W. E., Rototawai, Featherston.
BUICK, W. H., Masterton.
BATCHELAR, J. O., Palmerston North.
BOOTH, W. H., Middle Run, Carterton.
BATLEY, E. RUI, Moawhanga.
BAKER, W. E., Makino.
CLARK, T., Eskdale, Napier.
COBB, R., Kent Stud Farm, Mangaweka.
CHAMBERS, J. B., Te Mata, Hawke's Bay.
CORPE, J., Te Mara, Cunninghams.
CRESSWELL, E., Foxton.
CAMPBELL, E. A., Wiritoa, Wanganui.
COOK, G. L., Whakapuni, Hunterville.
DORSET, W. C., Kahikatea, Carterton.
DERMER, C. G. C., Cloverdene, Feilding.
DUNCAN & CAMPION, Ohirae, Fordell.
EGLINTON, H., Featherston.
ELLIS, W. A., York Farm, Marton.
EGLINTON, H. B., Kaitawa, Pahiatua.
ELDER, H. R., Waimahoe, Waikanae.
EAGLE, E., junr., Carterton.
FRENCH, R. O., Cunninghams.
FULTON, H. V., Crawford Street, Dunedin.
FULTON, GEO., Te Kumu, Mangaonoho.
FIELD, J. C., Gisborne.
FANNIN, A. R., Hiwi, Taihape.
GRAY BROS., Fairburn, Masterton.

MEMBERS—Continued.

GIBSON, WM., Kiwitea.

GARDNER, D. H., Koputaroa.

HUTCHINSON, F., Rissington, Hawke's Bay.

HARDING, A., Siberia, Ashhurst.

HADFIELD, H. S., Lindale, Otaihanga, Wellington.

HARDING, A., Mangawhare, Auckland.

HAWKINS, H. S., Glencoe, Hamilton.

HUNT, THOS., Highfield, Wakefield, Nelson.

HUTCHINSON, F., junr., Rissington, Hawke's Bay.

HOLMS, J., Waimahaka, Invercargill.

HARDING, J. W., Mount Vernon, Waipukurau.

HEWITT, J. E., Rangitawhaka, Kohinui, Mangatainoka.

JACOB, W. F., Kiwitea, Feilding.

KEBBELL, J., Te Rauawa, Ohau, Manawatu.

KELLY, J. C., Tokomaru.

KNIGHT, J., Feilding.

KUMMER, F. W. H., Mauriceville.

KILGOUR, D. H., Kiwitea.

KENSINGTON, F., Kakatuma, Rewa.

LAWSON, F. C. S., Tuakau, Auckland.

LEVETT, C. A. J., Kiwitea, Feilding.

LANCE, H. P., Cunninghams.

LYNCH, O. P., Emerald Glen, Paikakariki.

MATTHEWS, A., Waiorongomai, Featherston.

MASON, MAURICE, Taheke, Pukehou, Hawke's Bay.

MONCKTON, O. M., Patutahi, Gisborne.

MUNGAVIN, P., Willowbank, Porirua.

MITCHELL, J., Brockhurst, Marton.

MATTHEWS, H. A., Waiorongomai, Featherston.

McHARDY, P. A., Beaulieu, Palmerston North.

McHARDY, L. H., Blackhead, Waipawa.

McGREGOR, W. J. A. Mount Linton, Southland.

McGREGOR BROS., Masterton.

McKINNON BROS., D. and A., Ohinewai, Waikato.

McKENZIE & LOVELOCK, Cliff Side, Palmerston North.

McLEAN BROS., Waituna West.

MACKENZIE, J. R., Clinton, Otago.

MACKENZIE, R. H., Braemore, Hilton, Havelock North.

NIX, W. J., Tauherenikau, Featherston.

OATES, J. G., Peach Grove, Carterton.

PEARCE, W. G., Colyton.

PEARCE, W. B. V., Oroua Bridge,

PERRY, J. L., Wainiro, Waimata Valley, Gisborne.

REYNOLDS, G. M., Sandown, Mangaheia, Gisborne.

RIDDIFORD, E. J., Lower Hutt, Wellington.

ROWLAND, D., Bloomfield, Longburn.

MEMBERS—Continued.

REID, J. F., Elderslie, Oamaru.
REID, W., Makino.
RICHARDSON, F., Waituna.
RAYNER, W. H., Masterton.
SHORT, E., Parorangi, Feilding.
SYKES, GEO. C., Masterton.
SLACK, MRS. B. H., Oroua Bridge.
STANDEN, S., Feilding.
SMITH BROS., Paikakariki.
SMITH BROS., Woodlands, Colyton.
SMITH, T. ALFRED, Colyton.
SMITH, W. E., Brunswick, Wanganui.
STUCKEY, J. & SON, Opaki, Wairarapa.
TANNER, R., Longburn.
TYER, A., 17, Hill Street, Wellington.
VOSS, MAX, Karere, Longburn.
WHEELER, G. C., Burford, Stanway.
WHEELER, G. A., Rangitane, Kawhatau, Mangaweka.
WILSON, E., Porirua Road, Johnsonville.
WILSON, J. G., Bulls.
WILLIS, T. R., Greatford.
WATSON, H. N., Fairfield, Onga Onga, Hawke's Bay.
WILKINSON, H. C., Hinaburn, Featherston.
WILKINSON, JAS., Executors of, 20 Pirie Street, Wellington.
WINDLEY, W., Porirua.
WAUGH, THOS., Kimbolton.
WHITE, JAMES F., Ohaura, Greymouth, Westland.

RULES.

1. The name of the Association is "THE NEW ZEALAND ROMNEY MARSH SHEEP BREEDERS' ASSOCIATION."

2. The Registered Office of the Association will be for the present in Feilding.

3. The objects for which the Association is established are :—

   (a) To promote the Pastoral Industry by the encouragement of the breeding of Kent or Romney Marsh sheep, and the maintainance of the purity of the breed.

   (b) The establishment and publication of a Flock Book of recognised and purebred sires which have been used, or ewes which have been bred from, and of such other Flock Books (if any) which the Council may think fit, and the annual registration of the pedigrees of such sheep as are proved to the satisfaction of the Council to be eligible for entry.

   (c) The periodical compilation and publication of a statement of transactions connected with the breed, such as particulars relating to shows, sales, and other transactions, with such other general information with reference to the breeding and management of sheep, and to sheep-farming, as the Council may think fit.

   (d) The organising and holding of shows or exhibitions of Kent or Romney Marsh sheep ; the raising or providing of funds for the expenses thereof ; the undertaking or managing of combined members' sheep sales ; and the appointing of auctioneers and agents for the conduct of such sales.

   (e) The arrangement of classes, and the donation or augmentation of prizes and awards of certificates of merit at various shows, and the appointment or recommendation of judges thereat.

   (f) The investigation of cases of doubtful or suspected pedigrees.

   (g) The undertaking of the arbitration upon and settlement of disputes and questions relating to or connected with Kent or Romney Marsh sheep and the breeding thereof, and for other subsidiary purposes.

   (h) To provide rooms and other facilities for holding or conducting meetings for the objects or work of the Association.

(*i*) To purchase, take on lease, hire, receive by way of gift, or otherwise acquire, and also to sell, let, or dispose of any real or personal property for the purposes of the Association, subject to the provisions of "The Agricultural and Pastoral Societies' Act, 1877."

(*k*) To borrow from time to time any moneys required for the purposes of the Association upon such security as may be determined.

(*l*) To promote information with reference to sheep-breeding, by lectures, discussions, books, correspondence, or otherwise, and, for any of the objects of this paragraph of Rule 3, to co-operate with any university or college, or professor or lecturer thereof.

(*m*) To consider all questions affecting the interests of sheep-breeders; to initiate and watch over, and, if necessary, to petition Parliament, or any other authority, or to promote deputations in relation to measures affecting sheep-breeders, and, where necessary, to procure changes of the law affecting sheep, and the promotion of improvements in the administration thereof.

(*n*) To do all such other lawful acts and things as may be deemed incidental or conducive to the attainment of the above objects or any of them.

4. The income and property of the Association, from whatever source derived, shall be applied solely towards the promotion of the objects of the Association as set forth in these rules, and no part thereof shall be paid or transferred directly or indirectly by way of dividend, bonus, or otherwise howsoever, by way of profit to the members of the Association or any of them. Provided that nothing herein contained shall prevent the payment in good faith of remuneration to any salaried officer, nor the award of any prize given by the Association to any member, or prevent the *bona fide* borrowing of money by the Association from any member thereof at any rate of interest not exceeding 7 per centum per annum. No member shall derive any pecuniary gain, except as a salaried officer, from any of the property or operations of the Society.

5. If upon the dissolution of this Association there remains, after the satisfaction of all its debts and liabilities, any property whatsoever, the same shall not be paid to or distributed among the members of the Association, but shall be given or transferred to some other institution or institutions having objects similiar to the objects of the Association, to be determined by the members of the Association, at or before the time of dissolution, or in default thereof by such Judge of the Supreme Court as may have or acquire jurisdiction in the matter.

6. It is declared, for the purpose of the incorporation, that the number of members of this Association shall not be limited, but must not be less than fifty.

7. These rules shall be construed with reference to "The Agricultural and Pastoral Societies' Act, 1877," and any regulations made thereunder, and terms used in these rules shall be taken as having the same respective meaning as they have when used in that Act.

8. The management of the business of the Association shall be vested in a Council, consisting of fifteen members, to be elected from and by the members of the Association in a manner hereinafter provided.

9. All Romney breeders appearing in Vol. III. of the New Zealand Flock Book shall be invited to join and transfer their flocks to the New Zealand Romney Marsh Sheep Breeders' Association.

10. Any person taking an interest in Romney Marsh sheep who shall signify to the Council his desire to become a member, and who shall be proposed by some existing member of the Association, seconded by another member of the Association, shall, on election by a majority of the members present at a Council or General Meeting of the Association, be admitted as a member and entered on the register of members. Any member may at any time retire from the Association by giving notice in writing to that effect to the Secretary, provided that if an annual member he shall be liable and be required to pay any subscription or other payment which may be due from him and unpaid at the date of his retirement. Every member shall be either a life member or an annual member. A life member shall pay on admission a subscription of 10 guineas or such other sum as the Council may from time to time prescribe. An ordinary member shall pay on admission an entrance fee of one pound and an annual subscription of one pound, or such other sum as the Council may from time to time prescribe. Annual subscriptions shall be payable in advance on the 1st day of January in each year.

11. The Council may at any time elect as an honorary member of the Association any person who they consider to have advanced, or to be calculated to advance, the objects of the Association. Honorary members shall not be required to make any payments, and they shall not be eligible to hold office or to vote at any meeting of the Association.

12. Any member of the Association or of the Council who shall fail in the observance of any lawful rule, regulation, or by-law made by the Council, or whose conduct in any respect shall be, in the opinion of the Council, derogatory to the character or

prejudicial to the interests of the Association, may be removed from the Association (and if a member of the Council, from the Council) by a resolution to that effect passed by a majority of at least three-fourths of such members as shall be present and vote at a Special or Ordinary General Meeting, of which not less than fourteen days' previous notice, specifying the intention to propose such resolution, shall have been sent to all the members of the Association.

13. At the first General Meeting and at the Annual General Meeting in every subsequent year a President and Vice-President, Treasurer, and two Auditors shall be elected for the ensuing year. For this purpose the names of such persons as the Council shall think fit to serve as President, Vice-President, Treasurer, and Auditors shall be submitted to the meeting. Any other person may be proposed at such meeting, and the decision of the majority, taken by ballot, shall prevail.

14. The Council of the Association, until the first General Meeting, shall consist of all the subscribers, being members of the Association at the date of its formation. At the first General Meeting a list of all the persons being members shall be submitted to the meeting, from whom shall be elected fifteen members (including the President, Vice-President, and Treasurer) to constitute the Council.

15. At the first Annual General Meeting, and at the Annual General Meeting in every subsequent year, four of the members of the Council shall retire from office. The members to retire in each year shall be the members who have been longest in office since their last election. As between members of equal seniority the members to retire shall (unless such members of equal seniority shall agree among themselves) be selected from among them by ballot or in such other manner as the Council shall from time to time determine.

16. For the purpose of determining the seniority of the first and subsequent Councilmen, the Council shall, as soon as may be after the incorporation of the Association, arrange the first Councilmen in a list determining the order in which their names shall stand by ballot, or in such other manner as the Council shall think proper.

17. At the Annual General Meeting in each year in which members of the Council retire, the Association shall supply the vacancies arising from the retirement appointed to take place at such Annual General Meeting, by electing a like number of persons to be Councilmen. A retiring Councilman shall, in all cases, be eligible for re-election.

18. With regard to Councilmen, the mode of their election shall be as follows : The Council shall, with the notice convening the Annual General Meeting, cause to be sent to every member

the list containing the names of the four retiring members of the Council, and such other names as may have been proposed for election, such names to be printed in a form of voting paper to be approved by the Council. Such voting paper shall be filled up and returned to the Council by a date to be fixed by the Council, and must retain neither more nor less than **four** uncancelled names, and the Council shall appoint two Scrutineers to inspect votes, and report in writing to the meeting the names of the four persons having the highest number of votes. Such persons shall then be declared elected by the chairman of the meeting, and shall be deemed members of the Council, after which the voting list shall be forthwith destroyed by the Scrutineers. In the case of equality of votes, the names of the candidates who have an equal number of votes shall be submitted to the meeting, and a second ballot shall be taken at the meeting. A member may propose the name of any member or members for election on the Council by submitting the same to the Secretary not less than one month before the Annual General Meeting. Names so submitted shall be included on the list printed on the voting papers.

19. Any vacancies which may occur in the office of President, Vice-President, or Treasurer, or in the Council, between the Annual General Meetings, may be filled up by the Council. But any person so elected by the Council shall retain his office so long only as the member in whose place he is appointed would have retained it if no vacancy had occurred, and in the list of the Council mentioned in Rule 16 his name shall be placed where stood the name of the member in whose place he is appointed.

20. The Council shall be deemed to be duly constituted and shall continue to possess all the powers hereinafter stated, notwithstanding any vacancies in its body, but such vacancies shall be filled up as soon as, in the judgment of the Council, possible and expedient, in the manner pointed out in the preceding paragraph.

21. In all meetings of the Council one-third shall be a quorum.

22. The Council shall exercise all the powers and perform all the duties for which the Association has been established; and shall have full power to do all such things as may be incidental or conducive to the attainment of the objects of the Association mentioned in these rules.

23. In particular, but without prejudice to the last preceding article, and subject to the provisions of these rules, the Council shall have and may exercise and perform the following powers and duties :—

(*a*) They may from time to time convene and hold General Meetings of the Association.

(*b*) They may from time to time make, alter, and rescind rules, regulations, and by-laws, for registering the names and addresses of members ; for conducting the business and carrying out the objects of the Association, and for conducting the business of the Council, and they may impose fines for breach of any such rules, regulations, and by-laws. Provided nevertheless that no resolution of the Council shall be varied or rescinded, unless the resolution varying or rescinding the same be passed at one meeting of the Council and confirmed at a subsequent meeting.

(*c*) They may delegate any of their powers or duties (except the appointment and removal of members of their body, and of the Association) to Committees of any number (not less than three) of members of their body, and from time to time make, alter, and rescind regulations, rules, and by-laws for conducting the business delegated to such Committee.

(*d*) They may acquire for the Association any pedigrees or publications, with the copyright therein respectively (if any), the possession of which they may deem likely to be in any way advantageous to the Association ; and may establish any new publications devoted to or bearing on any object of the Association. The copyright of every publication acquired or established by or on behalf of the Association shall be vested in the Association.

(*e*) They may continue any contract with the printers or publishers of any publication acquired by the Association that may be subsisting at the date of such acquisition, and may enter into and make any new or altered contracts, or arrangements with the same, or any other printers and publishers, for the printing, publication, distribution, sale or management of such acquired publication, or of any publication whatsoever of the Association.

(*f*) They may from time to time regulate the nature, form, and contents of, and also the terms and conditions as to entries in, and also the time and mode and terms of issue of, any publication of the Association, and all arrangements and details connected therewith, and in particular they shall have power, so far as they may not be fettered by any subsisting contract or engagement, from time to time, to fix and alter the price of any publication of the

Association, and the charges to be levied for the insertion therein of entries relating to Romney Marsh sheep and other matters.

(g) They may accept annual or other subscriptions of money from members of the Association, or any other person or persons in payment for any publication of the Association, and they may make arrangements for supplying any such publication to any member or other person during his life, or for any other period, on the terms of receiving a lump sum of money in advance, or on such other terms as they may think fit.

(h) They may, subject to the provisions of " The Agricultural and Pastoral Societies' Act, 1877," purchase, hire, or take on lease for the purpose of the Association, any lands, tenements, houses, or parts of houses, and chattels, and they may sell, let and dispose of the same when and as they may think fit.

(i) They may from time to time appoint, employ and remove a Secretary of the Association, Editor or Editors of any publication of the Association, or any other officers, clerks and servants, at such salaries and wages respectively, and with such respective duties and spheres of employment, and generally, upon such terms as they may think fit.

(k) They may borrow money for the purposes of the Association at any rate of interest not exceeding 7 per cent. per annum, and may give security for any such moneys upon any property of the Association.

(l) They may place any moneys of the Association not required for immediate use upon deposit at interest at some bank, and they may invest any such moneys and generally deal with the property of the Association in such manner as they may think fit, and may for the time being be empowered to do under these rules.

24. The funds of the Association shall be applied as follows, namely :—

(a) In payment of the legal and other preliminary expenses incurred in the formation of the Association.

(b) In payment of the current expenses and other disbursements of the Council in the conduct of the business of the Association, or in relation thereto.

(c) In payment of the salaries and wages of the Secretary, Editor or Editors, and other officers, clerks and servants for the time being of the Association.

(d) In defraying all expenses of or connected with the printing, publication, sale and distribution of the publications of the Association.

(e) In paying the purchase-money or rent of any house, lands, goods, chattels, or effects purchased or hired by the Association, or any repairs or other outgoings in respect of such premises, and to paying for any other property required by the Association.

(f) In payment of the interest and repayment of the principal of any moneys borrowed by the Association, or in constituting a reserve fund to meet future contingencies, or in reducing the price charged for any publication of the Association, or generally upon or for any object or purpose expressed or implied by the rules; but the Council shall have power (subject always to the provisions of Rule 4, and to subsisting engagements) from time to time to vary the application of the funds of the Association in such a manner as they shall think fit.

25. The Council shall convene and hold a General Meeting of the Association within six months after the incorporation of the Association, and at such a place as they shall appoint. They shall convene and hold Annual General Meetings of the Association in such month, and on such day, and at such place, as they shall from time to time appoint.

26. A Special General Meeting of all members of the Association shall be held at such time and place as may be decided upon by the Council, for the purpose of making and adopting rules of the Association, or amending the same, and of transacting any general business (notice of which has been given) other than that to be transacted at the Annual Meeting. Fourteen days' notice shall be given of all meetings.

27. The Council may, whenever they think fit, and they shall, upon a requisition made in writing by not less than one-tenth in number of the members of the Association, convene a Special General Meeting.

28. Any requisition made by members shall express the object of the meeting proposed to be called, and shall be left at the registered office of the Association.

29. Upon receipt of such requisition, the Council shall forthwith proceed to convene a Special General Meeting. If they do not proceed to convene the same within twenty days from the date of the requisition, the requisitionists may themselves convene a Special General Meeting.

30. True accounts shall be kept of the sums of money received and expended by the Association, and the matters in respect of which such receipts and expenditure takes place, and of the assets,

credits, and liabilities of the Association in books of accounts, which shall be kept at the registered office of the Association, or at such other place or places as the Association thinks fit. At the first General Meeting, and subsequently at the Annual General Meeting in every year, two Auditors shall be appointed by resolution, who shall hold office till the next Annual General Meeting, but shall be eligible for re-election. The Auditors shall audit the accounts of the Association, previously to the Annual General Meeting, and a statement showing the financial position of the Association, and examined and certified by the Auditors, shall be sent to each member, and laid before every Annual General Meeting.

31. Any meeting may be adjourned as the members present thereat shall resolve.

32. The President of the Association, or in his absence the Vice-President, or in their absence any member then elected for the purpose by the members present, shall take the chair at all General Meetings.

33. All questions and matters brought before General Meetings, except the election of members of Council, which shall be decided as hereinbefore provided, shall be decided by a majority of votes of the members present, each member having one vote, and in the case of an equality of votes the Chairman of the meeting shall have a second or casting vote in addition to his vote as a member.

34. At General Meetings of the Association one-tenth part of the membership shall form a quorum.

35. The minute Books of the Association shall be open to the inspection of members at all reasonable times.

36. A notice may be served by or on behalf of the Association upon a member personally, or by sending through the post in a prepaid letter addressed to the member at his registered place of abode.

37. The Secretary shall keep the books of the Association and conduct the correspondence, attend all meetings, and receive members' subscriptions, and enter up the minutes of all meetings, and shall pay all moneys received for or on account of the Association to the Treasurer within seven days from the receipt of same.

38. The Treasurer shall receive all moneys from the Secretary, and bank the same to the credit of the Association, and shall sign all accounts of the Association.

39. The Association shall have a common seal bearing the words : " NEW ZEALAND ROMNEY MARSH SHEEP BREEDERS' ASSOCIATION," which seal shall be fixed to all deeds and documents

required to be sealed, and to all notices required to be authenticated. Such seal when required to be fixed to any deed or document shall be fixed thereto by two members of the Council of the Association in the presence of the Secretary, and such seal shall be kept at the registered office of the Association in the custody of the Secretary.

40. Sheep shall be eligible for entry in the New Zealand Romney Marsh Flock Book when transferred from a flock registered therein, or imported Romney sheep when accompanied by a certificate from its recognised Flock Book Society.

41. Imported sheep will be admitted to the New Zealand Romney Marsh Flock Book only on production of the certificate issued by its Flock Book Society, where such has been permanently established. If no such Society's certificate is forthcoming, its admission will be left to the discretion of the Council.

42. Both rams and ewes purchased from a registered flock must be transferred by the vendor, re-entered, and fees paid by the purchaser before the expiration of six months from the date of purchase. If application be made after the expiration of six months, double fees shall be charged. After the expiration of twelve months from date of purchase the application shall not be entertained.

43. No ram may be used in the registered flock of any member of this Association except from a flock also registered in this Association's Flock Book, or imported Romney rams, subject to Rule 41. Any infraction of this rule will entail exclusion of the progeny. All rams so used must be entered singly, and should be identified by label or tattoo.

44. All registered flock owners shall make their annual returns in each year, before some date to be fixed by the Council, and upon the Association's printed form. Any such owner not complying with this rule or in arrears with his subscription or fees, shall be omitted from the Flock Book, and shall cease to be a member of the Association, and shall not be re-admitted until such returns are supplied, together with double subscriptions and fees.

45. There shall be compiled and published annually by the Association in its Official Flock Book the breeding returns and all new flocks accepted by the Council ; deleting therefrom those flocks which have been dispersed, or whose owners have omitted to make their annual returns, or who have not paid their fees and annual subscriptions.

46. Such publications shall be made as soon after the date referred to in Rule 44 in each year as practicable, and under the direction of the Executive Committee of the Council, and shall give the number of ewes tupped, and the number of their lambs marked during the past year.

47. The said Council may appoint one or more competent men in such districts as they may consider necessary, who shall be an Inspection Committee, and shall inspect and count flocks whenever instructed by the Council. Should these Inspectors consider any such flock not fairly representative of its kind, the Council may expunge the said flock from the Flock Book.

48. The Council, when it thinks fit, may publish particulars of sale.

49. A registered flock owner, on production of a written statement that he is exporting a portion of his registered flock, may demand an export certificate from the Council, which shall be signed by the owner and endorsed by the President and the Secretary of the Association. In the case of stud rams, they shall be entered singly in the Association's next volume if not already entered, and the breeder shall take the necessary steps to identify each animal, either by label or tattoo, or by some other means approved of by the Council.

Fees for granting the foregoing export certificates shall be as follows, payable by the vendors :—

	s.	d.
For every ram or ram lamb ..	2	6 each
For every ewe or ewe lamb ..	1	0 each

50. If any inaccuracy be discovered relating to any entry, such entry may be cancelled by the Council, and the flock may be expunged from the Association's Flock Book. Should it at any time be proved that any entry is based upon any fraudulent statement or history, such entry shall be cancelled in like manner, and the flock expunged from the Flock Book.

51. The following fees shall be paid by members of this Association when entering their flocks : Transfer of sheep already registered, per 100 ewes or part thereof, £1 ; registration of single sheep 1/- each. Flocks on separate properties must make separate entries. The Council may, if it find it necessary, alter the membership, annual subscription, entrance, and other fees, to such a sum as may be required to meet expenses.

52. Notwithstanding anything to the contrary contained in any preceding Rule, flocks of indubitable purity may, for the present, be admitted to the New Zealand Romney Marsh Flock Book, subject to the following restrictions :—

(a) The applicant, in accordance with the rule framed when the New Zealand Sheep Breeders' Association was started, shall prove to the Council that his ewes in 1880 were admittedly purebred ; and further furnish a full statement as to rams used and ewes bought since then.

(*b*) The flock will be inspected and counted by one or more competent men selected by the Council, and from a distant district.

(*c*) That such applicant shall make an affidavit before a J.P. to the effect that his statement is correct, and shall pay all expenses in connection with such inspection, etc., an entrance fee of £3, and double fees for registration.

(*d*) That such application shall be advertised at least thrice with a view of enabling objections to be lodged ; and that further it shall be in the power of the Council to refuse registration in all such cases without assigning reasons for their action.

(*e*) Before any application is entertained, a sum of ten pounds (£10) must be deposited by the applicant to cover all fees and expenses in regard to such application, any balance unexpended to be refunded.

The foregoing Rules were adopted at a General Meeting called for that purpose, and held at Palmerston North, on the second day of November, one thousand nine hundred and four.

A. MATTHEWS, President.

NOTE—Communications to be addressed to—

The Secretary,

N.Z. Romney Marsh Sheep Breeders' Association,

P.O. Box 40, Feilding, N.Z.

DESCRIPTION

... OF A ...

Typical New Zealand Romney Marsh Sheep.

Drawn up by the Special Sub-Committee appointed for that purpose, and approved of at the Annual Meeting held at Masterton in February, 1907.

Head wide, level between ears, which should be large and thick, and covered with fine hair or down ; with good thick fore top, no horns, nor dark hair on the poll, which should be covered with wool.   Eyes should be large, bright, and prominent.   Face in Ewes full, and in Rams broad and masculine in appearance. Nose in all cases must be coal black.   Neck should be well set in at the shoulders, and strong and thick. Shoulders wide, well put in, and level with the back. Chest wide and deep. Back straight, with wide and flat loin.   Rump wide, long, and well turned. Tail set in almost level to the chine. Thighs well let down and developed. Legs should be short, with big bone, and large shapely feet of black horn.   The fleece should be of dense even quality and of a good decided staple from fore-top on the head to end of tail, and free from kemp.   The skin should be of a clean pink colour.

NOTICE TO MEMBERS.

The Council desires to remind the members of the Association that they can have photographs of their Flocks, or of any particular sheep, inserted in the Flock Book.

Full information as to costs can be obtained on application to

ERNEST J. WACKRILL,

Secretary,

P.O. Box 40, Feilding, N.Z.

INDEX.

OWNERS OF FLOCKS.

INDEX.

OWNERS OF FLOCKS—Continued.

OWNERS OF FLOCKS—Continued.

INDEX.

SEPARATE ENTRIES.

Rams.

RAMS—Continued.

SEPARATE ENTRIES.

Ewes.

No.	Name.		Address.		Page.
175—181	Buchanan, D. P.	...	Mayfield, Cunninghams	...	103
182—191	Ellis, W. A. ...	...	York Farm, Marton	...	104
159—174	Wheeler, G. C.	...	Burford, Stanway...	...	99

ROMNEY MARSH.

---o---

DIRECTIONS FOR THE READER.

Numbers in parentheses ( ) refer to Single Entries giving pedigrees, and may be found at the end of each volume of the Flock Book.

---o---

Mr. JAMES HOLMS,

WAIMAHAKA, INVERCARGILL.

Flock No. 1.

The Waimahaka flock was founded in June, 1876, with 35 ewes in lamb purchased at the late Mr. W. Tolmie's sale. The ewes were the progeny of ewes and rams purchased from the late Dr. Webster.

In 1905, 2 ewes bred by Mr. P. Mungavin (Flock 64), and in 1906, 4 ewes bred by Mr. A. Matthews (Flock 21), were added.

Previous to entry with this Association, rams used have been obtained from the late Mr. Ludlam, Dr. Webster, and Messrs. Matthews and W. B. Allen, and also imported from the flocks of Messrs. Rigden, Powell, and Millen, of Kent.

1906—Ewes retained from previous year	321	
Shearling Ewes home-bred	193	
Ewes purchased	6	
	Total	520
Lambs bred kept as Rams	280	
Ewe Lambs	308	

Sires –

Rigden's No. 33 (510)
Rigden's No. 75 (511)
Waimahaka No. 141 (512)
Waimahaka No. 142 (513)
Waimahaka No. 146 (514)
Waimahaka No. 149 (515)
Waimahaka No. 153 (515A)
Waimahaka No. 154 (515B)
Endurance (819)

Mr. J. F. REID,

ELDERSLIE, OAMARU.

Flock No. 5.

This flock was founded in 1879 by the purchase of the whole of Dr. Webster's flock. Since then importations have been made from Mr. H. Rigden, Kent.

Rams used previous to entry with this Association were bred by Messrs. M. Holmes, A. Matthews and Bealey Bros.; also imported rams bred by Messrs. J. S. Godwin, C. File, etc.

1906—Ewes retained from previous year	942
Shearling Ewes home-bred	285
Total	1227
Lambs bred kept as Rams	550
Ewe Lambs	555

Sires—

No. 255 (1041)	No. 288 (1025)
No. 256 (1042)	No. 280 (1032)
No. 263 (1043)	No. 281 (1027)
No. 271 (1044)	No. 412 (1014)
No. 264 (1045)	No. 276 (1035)
No. 257 (1046)	No. 381 (1024)
No. 265 (1047)	No. 309 (1018)
No. 270 (1048)	No. 417 (1011)
Elham No 31 (1036)	

Mrs. JAMES FULTON,

(Catherine Fulton, Widow of Hon. James Fulton.)

(MR. H. V. FULTON, MANAGER.)

RAVENSCLIFFE, WEST TAIERI, OTAGO.

Flock No. 6.

This flock was founded by the late Hon. James Fulton in the year 1869, with about 30 ewes and 1 ram purchased from the late Dr. Webster's flock.

In 1889, 6 ewes, bred by Mr. A. Matthews, were added to the flock, and in 1898 a further addition was made by the purchase of 6 ewes bred by Messrs. Gray Bros.

Previous to entry with this Association, imported rams, bred by Messrs. James Newport and H. Rigden, of Kent, and rams bred by Messrs. A. Matthews, M. Holmes, J. Holms and Gray Bros. have been used in this flock.

1906—Ewes retained from previous year	25
Ewes put to Rams	25
Lambs bred kept as Rams	12
Ewe Lambs	15

Sire—

Duke of Kent II. (1038)

The balance of this flock was exported to Australia.

Messrs. BEALEY BROS.,

HALDON, HORORATA, CANTERBURY,

Flock No. 9.

This flock is descended from sheep imported by Mr. Samuel Bealey and his brother, the late Mr. John Bealey, about the year 1867, and a further importation a couple of years later. They were from Messrs. Chittenden and Rigden, and, it is believed, another of the famous breeders in England. For a great many years the necessary changes of blood were obtainable within the flock, owing to the various strains being kept separate.

In 1893, 34 ewes from this flock were mated with Messrs. J. & A. D. McIlraith's ram " Masterton."

Previous to entry with this Association, imported rams, bred by Messrs. Godwin, Rigden, and Amos, of Kent, and a ram bred by the late Mr. A. C. Lang, and another from Mr. J. Reid, have been used in this flock.

1906—Ewes retained from previous year	152	**Sires—**
Shearling Ewes home-bred	73	Moses, imp. (369)
Total	225	Delphic Lad, imp. (370)
Lambs bred kept as Rams	87	Ashford Hero, imp. (368)
Ewe Lambs	98	Farmer's 26 of 1903, imp. (1050)

Mr. J. R. MACKENZIE,

POPOTUNOA, CLINTON, OTAGO.

Flock No. 19.

This flock was purchased from the late Mr. Donald Tolmie on 26th May, 1876. The ewes were in lamb at the time, and were from the flock of the late Dr. Webster.

This flock was added to in 1891 by 100 four-tooth ewes in lamb purchased from Captain Gardner, Birchwood; and again in 1892 by 50 aged ewes from the same flock.

In 1897, 2 ewes were imported, bred by Mr. W. Millen, of Kent.

Previous to entry with this Association, two imported rams, bred by Mr. W. Millen, and rams bred by Messrs. J. Holms, M. Holmes, Bealey, Watson, Shennan and Logan have been used in this flock.

1906—Ewes retained from previous year	276	**Sires—**
Shearling Ewes home-bred	170	Greenvale (703)
Total	446	Kelso (704)
		Jack (705)
Lambs bred kept as Rams	133	Conqueror (706)
Ewe Lambs	173	Glenkenich (707)
		Elderslie (708)
		Elderslie-Windsor (709)

Mr. ALFRED MATTHEWS,

WAIORONGOMAI, FEATHERSTON.

Flock No. 21.

The Waiorongomai flock was founded in 1875 by the purchase from Messrs. Barber of 83 ewes and lambs bred by Mr. Ludlam. In 1876, 156 more ewes and lambs were purchased from Messrs. Barber, also bred from Mr. Ludlam's flock. In March, 1881, 10 ewes were purchased from the late Mr. J. C. Boys, of Rangiora. In March, 1883, 2 ewes were imported from England. In 1894, 16 ewes were imported.

Previous to entry with this Association, rams used have been the imported sheep " Bismarck " and others from the flocks of Messrs. T. Cobb, H. Rigden, Chittenden, T. Giles, Neame, File, Wightwick, and Godwin, of Kent, and also from the flocks of Messrs. Ludlam, C. Pharazyn, E. J. Riddiford, J. C. Boys, Reid, Wheeler, Gray, J. Holms, M. Holmes, Mungavin, and Eglinton.

1906—Ewes retained from previous year	1750
Shearling Ewes home-bred	800
Total	2550
Lambs bred kept as Rams	906
Ewe Lambs	1365

Sires—

Rigden's No. 83, imp. (520)
Hazlewood 44th, imp. (522)
Windsor 83rd, imp. (523)
Westbroke No. 9, imp. (524)
Files Westbroke (104)
Hazlewood 42nd (110)
The Tenant (166)
No. 178 (757)
No. 63 (768)
No. 256 (790)
No. 23 (773)
No. 17 (785)
No. 227 (772)
No. 188 (754)
No. 97 (712)
No. 394 (767)
Windsor A 38 (913)
Burford No. 570 (830)
Burford No. 593 (833)

No. 38 (717)	No. 383 (743)
No. 508 (723)	No. 272 (775)
No. 221 (731)	No. 154 (749)
No. 129 (730)	No. 145 (748)
No. 40 (718)	No. 168 (751)
No. 19 (733)	No. 362 (770)
No. 311 (788)	No. 43 (779)
No. 83 (776)	No. 69 (780)
No. 78 (777)	No. 35 (783)
No. 93 (778)	No. 118 (786)
No. 141 (725)	No. 21 (784)
No. 207 (761)	No. 207 (787)
No. 65 (719)	No. 175 (1082)
No. 217 (762)	No. 172 (1084)
No. 206 (763)	No. 189 (1086)
No. 361 (771)	No. 125 (1088)
No. 215 (759)	No. 54 (1083)
No. 192 (756)	No. 235 (1085)
No. 146 (747)	No. 148 (1087)
No. 156 (745)	

Mr. ROBERT COBB,

KENT STUD FARM, MANGAWEKA.

Flock No. 22.

This flock was started in 1877 by the purchase from Mr. G. Hadfield, Moutoa, Foxton, of 80 ewes descended from Mr. Ludlam's flock. In 1883, 20 ewes were bought from the late Major Willis. In 1885, 12 ewe hoggets were procured from Mr. Eglinton, together with 3 older ewes, 6 ewes from Mr. West, Lower Hutt, and 10 ewes from Mr. T. Mitchell, Moutoa, Foxton.

Previous to entry with the Association, rams used were from the flocks of Messrs. G. Hadfield, E. Barber, Eglinton, and the owner ; also the late Mr. T. Cobb, and Messrs. Millen, Neame, Chittenden, and Gambrill, of Kent, England.

In 1897, after selecting 200 ewes and 80 ewe lambs, Mr. Cobb sold his farm, and the balance of the flock was dispersed.

1906—Ewes retained from previous year 380
Shearling Ewes home-bred 150
Total 530
Lambs bred kept as Rams 200
Ewe Lambs 300

Sires—
Perfection II. (694)
Symmetry (695)
Royal (696)
Model III. (791)
Jumbo X. (792)
Symmetry II. (793)
Buffalo Bill VI. (1126)
Active V. (1127)
Jumbo XI. (1128)

EXECUTORS of the late Mr. HENRY EGLINTON,

ORIENTAL BAY, WELLINGTON.

Flock No. 23.

The Whare Kauhau flock is descended from ewes (passing through the hands of Messrs. Ludlam, J. Dick, and Campbell) imported from England in 1857 by Mr. Leonard Young. In 1878, Mr. Eglinton purchased 20 ewes from Mr. Campbell; in 1886, 35 from Mr. Dick and 9 from Mr. T. P. Allen ; in 1883, 2 imported ewes bred by Mr. Thomas Cobb, Ivy Church, Kent.

Previous to entry with this Association, rams used have been from the flocks of Messrs. T. Finn and T. Neame, of Kent, England, and Messrs. Matthews, E. J. Riddiford, C. Pharazyn, G. Wheeler, Bealey Bros., and T. P. Allen.

1906—Ewes retained from previous year 384
Shearling Ewes home-bred 216
Total 600
Lambs bred kept as Rams 250
Ewe Lambs 318

Sires—
Whare Kauhau (401)
Whare Kauhau (402)
Whare Kauhau (404)
Whare Kauhau (405)
No. 85 (1129)   No. 93 (1132)
No. 88 (1130)   No. 99 (1133)
No. 90 (1131)

Mr. GEORGE WHEELER,

BURFORD, STANWAY, HALCOMBE.

Flock No. 24.

The Burford flock was founded in 1882 by the purchase of 105 ewes from the flock of the late Major Willis, of Woodendean, and the following year two grandsons of the imported ram "Colonel" were obtained. In 1887, 20 ewes were obtained from Mr. Reid, of Elderslie. In 1893, 10 ewes were imported from the flock of Mr. F. Neame, and in 1903, 2 ewes were also imported from the flock of Mr. J. S. S. Godwin, of Kent. In 1906, 2 ewes were imported from the flock of Mr. C. File, Elham, Kent.

Previous to entry with this Association, rams used in this flock have been obtained from Mrs. Willis, Messrs. Dorset, Cobb, Braithwaite, and Matthews, and also imported from the flocks of Messrs. File, Neame, and Godwin.

1906—Ewes retained from previous year	520
Shearling Ewes home-bred	210
Total	730
Lambs bred kept as Rams	280
Ewe Lambs	290

Sires—
Imperialist (794)
Socialist (795)
Freeholder (797)
Leaseholder (798)
Copyholder (799)
Landlord (810)
Godfrey (811)
Watchman 2nd (828)
Watchman (414)
Waterloo (413)
Burford No. 573 (831)
Burford No. 227 (1163)

Mr. WALTER C. DORSET,

KAHIKATEA, CARTERTON.

Flock No 25.

The Kahikatea flock is descended from the Manaia flock, which was started in 1861 by the late Mr. W. H. Donald, Manaia.

Previous to entry with this Association, rams have been obtained from Messrs. Pharazyn, Cobb, Reid, Eglinton, Riddiford, Sykes, Wheeler, W. B. Allen, J. Holms, and Batchelar.

1906—Ewes retained from previous year	200
Shearling Ewes home-bred	80
Total	280
Lambs bred kept as Rams	171
Ewe Lambs	202

Sires—
E.W.D. 022 (1164)
Kiwiteu 2nd (595)
Batchelar (837)
Waiwetu (580)

Mr. E. J. RIDDIFORD,

LOWER HUTT.

Flock No. 27.

This flock was founded in 1869 by the purchase of 50 ewes from Mr. Ludlam, Lower Hutt. In 1879, 15 ewes were purchased from Mr. Lyall, of Victoria, descended from stock imported from Mr. Phillpott, of Kent, England.

Previous to entry with this Association, rams were obtained from Messrs. Ludlam (including the imported ram " Colonel "), Reid, Matthews, M. Holmes, and Eglinton.

1906—Ewes retained from previous year	318	**Sires—**
Shearling Ewes home-bred	138	Lord Roberts (946)
	Total 456	Lord Roberts 11th (1166)
Lambs bred kept as Rams	258	Lord Roberts 28th (1167)
Ewe Lambs	250	Lord Roberts 33rd (1168)
		Lord Roberts 8th (1169)
		Lord Roberts 20th (1170)
		Fern Grove (432)

Mr. THOS. P. ALLEN,

WALMER FARM, HUTT.

Flock No. 35.

The Walmer Farm flock was started in 1884, when 21 ewes and 5 lambs were purchased from Mr. H. Campbell, of Wainui-o-mata. Mr. Campbell's flock was descended from 2 ewes from imported stock and the imported ram " Mousey," purchased in 1875 from Mr. A. Ludlam.

Previous to entry with this Association, rams bred by Messrs. Grace, Wheeler, Matthews, Dorset, Harding, Bidwill, W. B. Allen, Reid, J. Holms, and Bealey have been used in this flock.

1906—Ewes retained from previous year	22	**Sires—**
Shearling Ewes home-bred	20	No. 99 (589)
	Total 42	Handsome Jack (1171)
Lambs bred kept as Rams	21	Walmer 1. (1172)
Ewe Lambs	25	

Mr. W. B. ALLEN,

CLAREVILLE, WAIRARAPA.

Flock No. 36.

The Dealwood flock is descended from 15 ewes and 6 ewe lambs purchased in 1880 from Mr. H. Campbell, of Wainui-o-mata, who, in 1875, purchased from the late Mr. A. Ludlam, of the Hutt, 13 ewes. In 1894, 2 ewes bred by Messrs. G. W. Finn and Godwin, also a ram bred by Mr. G. W. Finn, were imported.

Previous to entry with this Association, rams have also been used from the following breeders : Messrs. Dick, Dorset, Matthews, T. P. Allen, Riddiford, Yorke, Reid, Mungavin, Gray, and J. Holms.

1906—Ewes retained from previous year	181		**Sires—**
Shearling Ewes home-bred		70	Waiorongomai No. 120 (1173)
	Total	251	Waiorongomai No. 334 (1174)
Lambs bred kept as Rams	116		Haldon (586)
Ewe Lambs	124		

Messrs. GRAY BROS.,

FAIRBURN, MASTERTON.

Flock No. 38.

The Fairburn flock was started in 1882 with 40 ewes purchased from Mr. Rhodes Donald, of Manaia, Masterton. In 1885, 3 ewes were purchased from Mr. J. Reid, of Elderslie.

Previous to entry with this Association, rams bred by Messrs. Bidwill, Reid, Matthews, Dorset, Cobb, Wheeler, Sykes, W. B. Allen, Fulton, Mungavin, and Eglinton have been used in this flock.

1906—Ewes retained from previous year	450		**Sires—**
Shearling Ewes home-bred		200	Elderslie 460 (205)
			Duke of Kent 4th (590A)
	Total	650	Parorangi B (590B)
Lambs bred kept as Rams	300		Parorangi C (590C)
Ewe Lambs	300		Parorangi D (590D)
			Elderslie 1902 (841)
			E.W.D. 9 (842)
			E.W.D. (839)
			Hororata 1st (1175)
			Hororata 2nd (1176)
			Hororata 3rd (1177)

Mr. W. J. NIX,

TAUHERENIKAU, FEATHERSTON.

Flock No. 39.

This flock is descended from 4 ewes and 1 ram, " Young Monarch," purchased in 1880 from Mr. J. C. Boys, of Canterbury. Ten ewes were also purchased from the Trustees of the late Mr. R. Barton's Estate, Fernside.

Previous to entry with this Association, rams from the flocks of Messrs. Reid, Monckton, Wheeler, Gray, Matthews, Cobb, Lang, Dorset, Pharazyn, Bealey, and W. B. Allen, and an imported ram bred by Mr. F. Neame, of Kent, have been used in this flock.

			Sires—	
1906—Ewes retained from previous year	340			
Shearling Ewes home-bred	80		W.J.N. No. 1 (843)	
			W.J.N. No. 2 (844)	
	Total	420	W.J.N. No. 3 (845)	
Lambs bred kept as Rams	180		(151)	(644)
Ewe Lambs	230		(215)	(681)

Mr. JOHN KEBBELL,

OHAU, MANAWATU.

Flock No. 40.

The Rauawa flock was founded in March, 1878 (at which time Mr. A. Braithwaite, of Waiwetu, was a partner), with a draft of 20 two-tooth ewes bred by Mr. Braithwaite out of ewes he had purchased from Mr. A. Ludlam, of the Hutt, at the dispersion of his flock, and by the imported ram " Bismarck."

Previous to entry with this Association, imported rams bred by Messrs. Rigden and Page, of Kent, and rams bred by Mrs. Willis, Messrs. Fulton, Matthews, Dorset, Gray, and Wheeler have been used in this flock.

			Sires—
1906—Ewes retained from previous year	181		
Shearling Ewes home-bred	52		Reid's No. 278 (592D)
			Reid's No. 253 (592B)
	Total	233	Reid's No. 252 (592A)
Lambs bred kept as Rams	136		Reid's No. 254 (592C)
Ewe Lambs	120		

Mr. DAVID BUICK,

CLOVERLEA, PALMERSTON NORTH.

———

Flock No. 42.

The Cloverlea flock is descended from 17 ewes purchased in 1889 from Mr. T. P. Allen, Hutt (Flock No. 35). These were the progeny of ewes purchased in 1875 by Mr. Campbell, of Wainui-o-mata, from Mr. A. Ludlam. The Cloverlea flock was added to in 1892 by the purchase from Mr. J. C. Yorke (on his departure for England) of 70 ewes ; these were descended from the flock of Mr. J. C. Boys, of Rangiora. In 1904, 5 ewes were purchased at Mr. Bridge's dispersion sale (Flock No. 56).

Previous to entry with this Association, rams used in this flock were bred by Messrs. York, Cobb, T. P. Allen, and Rowland.

1906—Ewes retained from previous year	516
Shearling Ewes home-bred	300
Total	816
Lambs bred kept as Rams	390
Ewe Lambs	370

Sires—

Fairfield-Windsor No. 161 (847)
Fairfield-Windsor No. 149 (848)
Fairfield-Windsor No. 167 (849)
Fairfield-Windsor No. 164 (850)
Fairfield-Windsor No. 156 (851)
Fairfield-Windsor No. 158 (852)
Cloverlea-Walmer No. 1 (1178)
Cloverlea-Walmer No. 2 (1179)
Cloverlea-Walmer No. 3 (1180)
Cloverlea-Walmer No. 4 (1181)
Cloverlea-Walmer No. 5 (1182)
Cloverlea No. 23 (853)
Cloverlea No. 24 (854)
Cloverlea No. 25 (855)
Cloverlea No. 26 (856)

Mr. W. F. JACOB,

KIWITEA, NEAR FEILDING.

Flock No. 43.

This flock is descended from ewes purchased in 1892 from Mr. J. C. Yorke (late of Inaha, Hawera), whose ewes were purchased from Mr. J. C. Boys, of Rangiora, in 1880. In 1905, 16 ewes with lambs, purchased at the dispersion sale of Mr. J. O. Batchelar (Flock 45) were added.

Previous to entry with this Association, rams used were bred by Mrs. Willis and Messrs. Wheeler, Matthews, Nixon, and Bealey ; also, imported rams bred by Messrs. Arthur Finn and Godwin, of Kent.

			Sires—
1906—Ewes retained from previous year	176		
Shearling Ewes home-bred	90		Westbroke 45 (240)
	Total	266	
Lambs bred kept as Rams	121		Westbroke 1 (302)
Ewe Lambs	172		Young Godwin (593)
			Pilgrim (857)
			Lyddite (858)
			Kiwitea 7th (862)
			Kiwitea 8th (863)

Mr. BERNARD CHAMBERS,

TE MATA, HAVELOCK NORTH.

Flock No 46.

The Te Mata flock was founded in February, 1893, by the purchase of 150 ewes from Mr. J. Reid, of Elderslie, and 100 ewes bred by Captain Gardner, of Birchwood, Southland.

In 1904, 9 ewes were purchased at Mr. Bridge's dispersion sale.

Previous to entry with this Association, rams bred by Messrs. Reid, Harding, Lang, Matthews, Wheeler, McKenzie, and Hutchinson have been used in this flock.

1906—Ewes retained from previous year	397	Sires—
Shearling Ewes home-bred	130	Windsor 50th, imp. (259)
Total	527	The Sovereign (94)
		Bombardier (79)
Lambs bred kept as Rams	244	Afridi II. (248)
Ewe Lambs	233	The Bishop (604)
		Nuncio (869)
		Hazlewood No. 15 (241)
		Hilton A (870)
		Model (298)
		Top Knot (300)

Mr. GEORGE R. SYKES,

MASTERTON.

Flock No. 48.

This flock was founded in 1894 by the purchase of 9 ewes from Mr. H. Eglinton (Flock No. 23), and a ram from Mr. A. Matthews (Flock No. 21).

Previous to entry with this Association, rams bred by Messrs. Matthews, Eglinton, Wheeler, Rowlands, Gray, and Bealey have been used in this flock.

1906—Ewes retained from previous year	135	Sires—
Shearling Ewes home-bred	60	Ready II. (871)
Total	195	Manaia (872)
		Manaia II. (873)
Lambs bred kept as Rams	95	
Ewe Lambs	110	

Mr. H. S. HADFIELD,

LINDALE, OTAIHANGA, WELLINGTON.

Flock No. 49.

The Lindale flock was founded in February, 1894, by the purchase of 50 ewes from the flock of Mr. J. W. Marshall, and in 1896, 37 ewes from the same flock were added.

Previous to entry with this Association, rams bred by Messrs. Wheeler, Bealey, Matthews, and Short have been used in this flock.

1906—Ewes retained from previous year	139	Sires—
Shearling Ewes home-bred	61	Hereward (463)
	Total 200	Lost Chord (565)
Lambs bred kept as Rams	81	Premier (557)
Ewe Lambs	97	Burford No. 421 (814)

Mr. DAVID ROWLAND,

BLOOMFIELD, LONGBURN.

Flock No. 50.

This flock was founded in 1877 by the purchase of 80 ewes bred by Mr. Ludlam. One imported ewe, with ewe lamb at foot, was added in 1894.

Previous to entry with this Association, rams have been obtained from Messrs. Akers, Boys, and Cobb, also imported rams from Messrs. Rigden and File, of Kent, England.

1906—Ewes retained from previous year	380	Sires—
Shearling Ewes home-bred	90	Windsor 3, imp. (255)
	Total 470	Bloomfield 6 (466)
Lambs bred kept as Rams	265	Bloomfield 7 (467)
Ewe Lambs	270	Bloomfield 8 (874)
		Bloomfield 9 (875)
		Bloomfield 10 (876)
		Bloomfield 11 (877)
		Bloomfield 12 (1197)
		Bloomfield 13 (1198)
		Bloomfield 14 (1199)
		Bloomfield 15 (1200)
		Bloomfield 16 (1201)
		Manuku (1202)

EXECUTORS of the Late Mr. JAS. WILKINSON,

(20, Pirie Street, Wellington.)

LAKE MEADOWS, FEATHERSTON.

Flock No. 57.

The Lake Meadows flock dates its origin in the purchase of 65 ewes from Messrs. Barber Bros., of the Lower Hutt, in 1879, being the progeny of ewes bred by the late Mr. A. Ludlam, and also a few ewes by the same breeder purchased from Mr. A. Matthews.

Previous to entry with this Association, rams bred by Messrs. Matthews and Pharazyn have been used in this flock.

			Sires—
1906—Ewes retained from previous year	628		
Shearling Ewes home-bred	333		Lake Meadows No. 16 (277)
			Lake Meadows No. 18 (279)
	Total	961	Lake Meadows No. 19 (606)
Lambs bred kept as Rams	65		Lake Meadows No. 20 (607)
Ewe Lambs	398		Lake Meadows No. 21 (608)
			Lake Meadows No. 22 (609)
			Lake Meadows No. 23 (610)
			Lake Meadows No. 24 (611)
			Lake Meadows No. 25 (612)
			Lake Meadows No. 26 (613)
			Mahoi 1 (614)
			Mahoi 2 (615)
			Mahoi 3 (616)
			Mahoi 4 (617)
			Mahoi 5 (618)
			Tahora 1 (1203)
			Tahora 2 (1204)
			Tahora 3 (1205)
			Tahora 4 (1206)

Mr. FRED C. S. LAWSON,

WAIRAMARAMA, TAUKAU, AUCKLAND.

Flock No. 62.

The Roto-o-rangi flock was purchased from Mr. A. Braithwaite in 1893, and became the property of Mr. Lawson in 1901.

In 1903, 30 ewes were purchased from the New Zealand Lands Association (formerly Flock No. 80) at their dispersion sale.

Previous to entry with this Association, rams bred by Messrs. Matthews, Barugh, and Hawkins have been used in this flock.

			Sires—
1906—Ewes retained from previous year	69		
Shearling Ewes home-bred	30		Ajax (878)
	Total	99	Vulcan (879)
Lambs bred kept as Rams	26		
Ewe Lambs	45		

Mr. ROBT. TANNER,

LONGBURN.

Flock No. 63.

This flock was founded in 1891 by the purchase, from Mr. D. Rowland (Flock No. 50), of 71 ewes. In 1893, 100 ewes were added from the same flock. In the same year 2 ewes were added from the flock of Mr. J. Reid (Flock No. 5).

Previous to entry with this Association, rams from the flocks of Messrs. Cobb, Wheeler, T. P. Allen, Reid, Gray, Dorset, Mungavin, W. B. Allen, and Batchelar have been used ; also 2 imported rams bred by Mr. File, belonging to Mr. D. Rowland.

1906—Ewes retained from previous year	355	Sires—
Shearling Ewes home-bred	175	Tiritea Duke (603)
		Admiral (168)
Total	530	Restorer (640)
Lambs bred kept as Rams	218	Sampson 2nd (477)
Ewe Lambs	292	Lord Elderslie 107 (880)
		Bloomfield 4th (1207)
		Windsor A 35 (464)

Mr. P. MUNGAVIN,

WILLOWBANK, PORIRUA.

Flock No. 64.

The Willowbank flock was started in 1892 by the purchase of 200 ewes from Mr. J. Kebbell, of Ohau (Flock No. 40), and Mr. F. A. Death, of Pahautanui (Flock No. 34). In 1896, 20 ewes were added from the flock of Mr. F. A. Death (Flock No. 34). In 1891, 8 ewes were purchased from Mr. T. P. Allen (Flock No. 35).

Previous to entry with this Association, an imported ram, bred by Mr. H. Rigden, Kent, and rams bred by Messrs. Matthews, Cobb, Rowland, Wheeler, Dorset, and Reid have been used in this flock.

1906—Ewes retained from previous year	165	Sires—
Shearling Ewes home-bred	60	Watchman 3rd (829)
		Sampson (1208)
Total	225	Tattoo No. 5 (1209)
Lambs bred kept as Rams	100	Tattoo No. 8 (1210)
Ewe Lambs	125	Porirua No. 15 (1211)

Mr. HERBERT S. HAWKINS,

GLENCOE, HAMILTON.

———

Flock No. 65.

This flock was started in 1892 by the purchase of 100 ewes from Mr. W. Akers (Flock No. 73).

These ewes were put to 2 rams bred by Mr. Edward Braithwaite. In 1893, 10 ewe lambs, bred by Mr. Edward Braithwaite, were added to this flock.

Previous to entry with this Association, rams bred by Messrs. Barugh, Cobb, Wheeler, and the New Zealand Lands Association have been used in this flock.

1906—Ewes retained from previous year	178	**Sires—**
Shearling Ewes home-bred	79	Corporal (151)
		Turk (152)
Total	257	Woodlands (482)
Lambs bred kept as Rams	139	Woodlands No. 1 (625)
Ewe Lambs	138	Glencoe No. 2 (294)
		Glencoe No. 10 (1212)
		Glencoe No. 11 (1213)
		Glencoe No. 12 (1214)
		Glencoe No. 13 (1215)
		Glencoe No. 14 (1216)

Messrs. F. HUTCHINSON & SON,

RISSINGTON, HAWKE'S BAY.

Flock No. 71.

The Rissington flock was started in 1891 by a purchase of 46 ewes, in part bred by Messrs. Bealey Bros., Canterbury, the remainder being progeny of same bred by Mr. S. Bolton. An addition was made in 1893 by a purchase at Mr. J. C. Yorke's clearance sale. Mr. Yorke obtained his first ewes from Mr. J. C. Boys, of Canterbury. A further small addition was made in 1897 of ewes bred as described in N.Z. Flock Book, 1898. In 1904, 30 ewe lambs were purchased at the Woodendean dispersion sale (Flock No. 26)

Previous to entry with this Association, imported rams bred by Messrs. Finn and Godwin, of Kent, and rams bred by Messrs. Cobb, Matthews, Yorke, and Gray have been used in this flock.

1906—Ewes retained from previous year	390	Sires—
Shearling Ewes home-bred	150	Cardinal V. (887)
		Cardinal VI. (1217)
Total	540	John Bull (485)
5 of these Ewes were sent to Burford		Sir Arthur (1222)
and tupped by—		Imperialist (794)
Lambs bred kept as Rams	120	No. 6 of 1903 (1218)
Ewe Lambs	250	No. 7 of 1903 (1219)
		No. 27 of 1903 (1220)
		No. 4 of 1903 (1221)
		Squire Finn (1223)
		Tonbridge (1224)

Mr. J. WALDEN HARDING,

MOUNT VERNON, WAIPUKURAU.

Flock No. 72.

The Mount Vernon flock was started in 1876, when rams were bought from the late Mr. Alfred Ludlam, of the Hutt, and ewes from Mr. Boys, Rangiora, Canterbury.

Previous to entry with this Association, rams used were purchased from the late Mr. Ludlam, Dr. Webster, Messrs. Holmes, T. P. Allen, Matthews, and Bealey.

1906—Ewes retained from previous year	133	Sires—
Shearling Ewes home-bred	28	Young Regent (Burford 132) (409)
		Bealey's Pride (699)
Total	161	Windsor 61st (500)
Lambs bred kept as Rams	38	
Ewe Lambs	38	

Mr. WILLIAM AKERS,

RIVERSDALE, PALMERSTON NORTH.

Flock No. 73.

This flock was founded in 1872 by the purchase of 80 ewes from the flock of the late Mr. Ludlam.   In 1883, 3 imported ewes, in lamb to a ram bred by Mr. Godwin, of Kent, were purchased from Mr. Cobb.

Previous to entry with this Association, imported rams bred by Messrs. Rigden and Finn, of Kent, and rams bred by Major Willis, Messrs. Holmes, Wilkin, and Dorset have been used in this flock.

1906—No returns.

ESTATE of the late Mr. R. HENRY MACKENZIE,

HILTON, HAVELOCK NORTH.

Flock No. 76.

This flock was founded in 1889 by the purchase of 230 ewes and hoggets given by Mr. John Harding, of Mount Vernon, to his son, Mr. W. R. Harding, and 45 ewes purchased from Mr. H. Eglinton (Flock No. 23) have been added.

Previous to entry with this Association, rams bred by Messrs. Bealey, Chambers, and Matthews have been used in this flock.

1906—			Sires—
Ewes retained from previous year		650	
Shearling Ewes home-bred		250	2 Reid's Elderslie
			2 Chambers' Te Mata
	Total	900	1 Wheeler
Lambs bred kept as Rams	149		9 Hilton
Ewe Lambs	244		Hilton 13 (900)
			Hilton 14 (901)

Mr. E. SHORT,

PARORANGI, WAITUNA WEST, WELLINGTON.

Flock No. 77.

This flock was founded in 1897 by the purchase of 150 ewes in lamb from Mr. R. Cobb (Flock No. 22), and in 1901, 36 ewes and 30 ewe lambs, and in 1902, 31 shearling ewes were purchased from the same breeder.

In 1905, 7 ewes and 2 ewe lambs were purchased at Mr. Batchelar's dispersion sale.

Previous to entry with this Association, two imported rams, bred by Mr. C. File, of Kent, and rams bred by Messrs. W. B. Allen, Gray, Dorset, Batchelar, Rowland, Cobb, and Wheeler have been used in this flock.

1906—Ewes retained from previous year	800	Sires—
Shearling Ewes home-bred	300	Earl Godwin (301)
		Favorite (330)
Total	1100	Romeo (332)
Lambs bred kept as Rams	580	Parorangi 2nd (342)
Ewe Lambs	590	Chancellor (346)
		Waiorongomai (384)
		Major 2nd (488)
		Batchelor 2nd (498)
		Windsor 60th (501)
		Buffalo Bill 5th (507)
		Young Hazlewood (594)
		Togo (601)
		Young Model (629)
		Nugget (630)
		Reserve (655)
		Record (663)
		Hazlewood (860)
		Young Westbroke (869)
		Prince of Wales (905)
		President (907)
		Windsor A 36 (911)
		Windsor A 37 (912)
		Nonsuch B 9 (921)
		The Marshal B 13 (922)
		Earl Roberts (946)
		Sandow E 6 (1272)
		Matchless (1273)

Messrs. McHARDY BROS.,

BEAULIEU, PALMERSTON NORTH.

Flock No. 79.

This flock was founded in 1901 by the purchase of 120 ewes and 36 ewe lambs from Mr. Wakelin, representing the balance of Flock No. 59, dispersed. In 1906, 9 ewes, bred by Mr. R. Cobb (Flock 22), were added.

Previous to entry with this Association, rams bred by Messrs. Dorset and Sykes have been used in this flock.

		Sires—
1906—Ewes retained from previous year	211	
Shearling Ewes home-bred	103	No. 908 (954)
Ewes purchased	9	No. 909 (955)
		No. 910 (956)
Total	323	No. 911 (957)
Lambs bred kept as Rams	196	Hororata 905 (664)
Ewe Lambs	196	Hororata 906 (665)

This flock is now divided—Mr. Leslie McHardy transfers 224 ewes and 98 ewe lambs to Blackhead, Waipawa, Hawke's Bay, and Mr. P. A. McHardy retains the same number at Beaulieu ; also purchases from Messrs. J. Stuckey & Son (Flock No. 164) 19 ewe lambs bred by them, and 17 ewes bred by Mr. G. Wheeler (Flock No. 24).

Mr. THOS. CLARK,

HEDGELEY, ESKDALE.

Flock No. 81.

This flock was founded in 1901 by the purchase of 100 ewes from Mr. G. Wheeler (Flock No. 24).

Previous to entry with this Association, rams bred by Messrs. Wheeler and Bridge have been used in this flock.

		Sires—
1906—Ewes retained from previous year	80	
Shearling Ewes home-bred	35	Zulu (101)
		Porirua No. 3 (671)
Total	115	Zulu No. 2 (967)
Lambs bred kept as Rams	42	
Ewe Lambs	42	

Mr. HERBERT N. WATSON,

FAIRFEILD, ONGA ONGA, HAWKE'S BAY.

Flock No. 82.

This flock was founded in 1900 by purchase from Messrs. Barron Bros., who, in 1897, bought the whole of Mr. Guthrie Smith's flock. The latter flock was started in 1870 by Mr. W. D. Lawrence with 2 imported rams from Sir H. Wild, and 25 ewes from Sir E. Stafford. Subsequently 20 ewes were purchased from Mr. Tolmie, 16 from Dr. Webster, and 42 from Mr. Boys.

Previous to entry with this Association, rams bred by Dr. Webster, Messrs. Ludlam, Bealey, Boys, Reid, Harding, Wheeler, and Monckton have been used in this flock.

In 1904 Mr. Watson took 64 ewes of the above flock from Patutahi, Gisborne, to Fairfield, and purchased 16 ewes, bred by Mr. H. H. Bridge (Flock No. 56), at the dispersion sale the same year.

In 1905, 1 ewe, bred by Mr. C. J. G. Hulkes, Kent, was imported.

1906—Ewes retained from previous year	104	Sires—
Shearling Ewes home-bred	25	Leader (812)
Ewes purchased	1	Westbroke No. 17 of 1904 (1298)
	Total 130	Too Late (422)
Lambs bred kept as Rams	71	
Ewe Lambs	80	

Mr. GEORGE W. BULL,

WAIKIEKIE, AUCKLAND.

Flock No. 83.

This flock was founded in April, 1903, by the purchase of 10 ewes in lamb from the New Zealand Lands Association, Limited (Flock No. 80). In 1905, 40 ewes were purchased from Messrs. F. Hutchinson & Son (Flock No 71).

1906—Ewes retained from previous year	45	Sire—
Shearling Ewes home-bred	5	Waikiekie No. 1 (673)
	Total 50	
Lambs bred kept as Rams	28	
Ewe Lambs	31	

Mr. E. J. WILSON,

PORIRUA ROAD, JOHNSONVILLE.

Flock No. 84.

This flock was founded in 1903 by the purchase of 50 ewes and 20 ewe lambs from Messrs. Jones Bros. (Flock No. 69), 31 ewe lambs from Mr. E. Short, which were bred by Mr. D. Rowland (Flock No. 50), and 16 ewes from Mr. T. P. Allen (Flock No. 35).

Previous to entry with this Association, rams bred by Messrs. Dorset, T. P. Allen, and Jones Bros. were used in this flock.

1906—Ewes retained from previous year	22	**Sires—**	
Shearling Ewes home-bred	26	120 (505)	
Total	48	121 (506)	
Lambs bred kept as Rams	20		
Ewe Lambs	35		

Mr. FRANK KENSINGTON,

RAKATUMA, REWA, NEAR WAITUNA WEST.

Flock No. 85.

This flock was founded in 1903 by the purchase, from Mr. E. Short, of 70 ewes bred by Mr. D. Rowland (Flock No. 50), with their progeny, 40 hoggets, and 48 ewe lambs. In 1905, 6 ewes with lambs were purchased from Mr. J. O. Batchelar (Flock No. 45), and in 1906, 25 ewes, purchased from Mr. C. E. Lucas (Flock No. 129), were added.

1906—Ewes retained from previous year	119	**Sires—**	
Shearling Ewes home-bred	50	Willowbank 2nd (454)	
Ewes Purchased	31	Rakatuma (661)	
Total	200	Rakatuma 3rd (972)	
		Rewa D7 (930)	
Lambs bred kept as Rams	90		
Ewe Lambs	80		

Mr. J. C. ALLEN,

ANNANDALE, PIAKO, AUCKLAND.

Flock No. 86.

This flock was founded in 1903 by the purchase of 30 ewes in lamb bred by the New Zealand Lands Association, Woodlands (Flock No. 80).

		Sire—
1906—Ewes retained from previous year	35	
Shearling Ewes home-bred	13	Gordon I. (674)
Total	48	
Lambs bred kept as Rams	25	
Ewe Lambs	27	

Dr. G. E. ANSON,

TE KUMU, MANGAONOHO.

Flock No. 87.

This flock was founded in 1904 by the purchase of 48 two-tooth maiden ewes at the dispersion sale of the Woodendean Flock (No. 26).

		Sire—
1906—Ewes retained from previous year	36	
Ewes put to Rams	36	Carfax (574)
Lambs bred kept as Rams	27	
Ewe Lambs	20	

Mr. WILLIAM BAKER,

MAKINO DOWNS, FEILDING.

Flock No. 88.

This flock is now dispersed.

Mr. JAMES BELL,

BRANDON HALL, BULLS.

Flock No. 89.

This flock is now dispersed.

Mr. H. P. BEYERS,

THE BROW, WAIPAWA.

Flock No. 90.

This flock was founded in 1904 by the purchase of 80 ewes from Mr. J. W. Harding's Mount Vernon Flock (No. 72). In 1905, a further purchase of 30 ewes from the same flock was added.

1906—Ewes retained from previous year	105	Sires—
Shearling Ewes home-bred	35	Fairfield (Flock No. 56)
Total	140	Hazlewood 15th (241)
Lambs bred kept as Rams	52	
Ewe Lambs	60	

Mr. DAVID PENRUDDOCKE BUCHANAN,

MAYFIELD, CUNNINGHAMS.

Flock No. 91.

This flock was founded in 1904 by the purchase of 15 pedigree ewes bred by Mr. G. Wheeler (Flock No. 24), see single entry of ewes; also 13 ewes bred by Mr. J. Holms (Flock No. 1), and 5 ewes at the dispersion sale of the Woodendean Flock (No. 26).

1906—Ewes retained from previous year	31	Sire—
Shearling Ewes home-bred	21	Harold (571)
Total	52	
Lambs bred kept as Rams	20	
Ewe Lambs	33	

Mr. JOSEPH CORPE,

TE MARA, CUNNINGHAMS.

Flock No. 92.

This flock was founded in 1904 by the purchase of 11 ewes at the dispersion sale of the Woodendean Flock (No. 26).

			Sire—
1906—Ewes retained from previous year	11		
Shearling Ewes home-bred	5		Kentishman (983)
		Total 16	
Lambs bred kept as Rams	10		
Ewe Lambs	7		

Mr. ARTHUR R. FANNIN,

HIWI, TAIHAPE.

Flock No. 93.

This flock was founded in 1904 by the purchase of 23 two-tooth ewes bred by Messrs. W. R. Perston & Son (Flock No. 61).

			Sires—
1906—Ewes retained from previous year	19		
Shearling Ewes home-bred	14		Dealwood 4th (1312)
		Total 33	Rover (985)
Lambs bred kept as Rams	19		
Ewe Lambs	21		

Mr. JAMES COPELAND FIELD,

HOMEBUSH, GISBORNE.

Flock No. 94.

This flock was founded in 1904 by the purchase of 74 ewes bred by Mr. H. N. Watson (Flock No. 82). In 1905 further purchases were added of 31 ewes bred by Mr. G. Wheeler (Flock No. 24), and 23 ewes from Messrs. Sproule & McCutchan, who obtained them from Mr. H. N. Watson (Flock No. 82). In 1906, 22 ewes bred by Gray Bros. were added.

			Sires—
1906—Ewes retained from previous year	117		
Shearling Ewes home-bred	37		Vice Regent (156)
Ewes purchased	21		Cardinal IV. (109)
		Total 175	Emperor (570)
Lambs bred kept as Rams	95		
Ewe Lambs	75		

Mr. RICHARD OTWAY FRENCH,

CUNNINGHAMS.

Flock No. 95.

This flock was founded in 1904 by the purchase of 5 ewes at the dispersion sale of the Woodendean Flock (No. 26).

1906—Ewes retained from previous year	5	Sire—
Shearling Ewes home-bred	1	Jacob 308 (600)
	Total 6	
Lambs bred kept as Rams	4	
Ewe Lambs	1	

Mr. JOHN EDWARD HEWITT,

RANGITAWHAKA, KOHINUI.

Flock No. 96.

The Rangitawhaka flock was founded in 1905 by the purchase of 30 ewes bred by Mr. J. Holms (Fiock No. 1).

1906—Ewes retained from previous year	27	Sire—
Ewes put to Rams	27	Kuroki (693)
Lambs bred kept as Rams	16	
Ewe Lambs	12	

Mr. JAMES KNIGHT,

FEILDING.

Flock No. 97.

This flock was founded in 1904 by the purchase of 40 shearling maiden ewes at the dispersion sale of the Woodendean Flock (No. 26). In 1905, 15 ewes with lambs were purchased at the dispersion sale of Mr. J. O. Batchelar (Flock No. 45), and 24 ewes bred by Mrs. Willis (Flock No. 26) were purchased from the executors of the late Mr. W. Baker.

1906—Ewes retained from previous year	76	Sire—
Ewes put to Rams	76	Massey (645)
Lambs bred kept as Rams	28	
Ewe Lambs	41	

Mr. HENRY P. LANCE,

BUCKLAND, CUNNINGHAMS.

Flock No. 98.

This flock was founded in 1904 by the purchase of 10 shearling ewes in lamb, bred by Mr. G. Wheeler (Flock No. 24).

1906—Ewes retained from previous year	10	**Sire—**
Shearling Ewes home-bred	6	Ratanui 10th (678)
Total	16	
Lambs bred kept as Rams	7	
Ewe Lambs	6	

Mr. OHAS. ARTHUR J. LEVETT,

RATANUI, KIWITEA.

Flock No. 99.

This flock was founded in 1904 by the purchase of 8 ewes bred by Mr. J. Holms (Flock No. 1), and 10 ewes in lamb, at the dispersion sale of the Woodendean Flock (No. 26).

1906—Ewes retained from previous year	15	**Sire—**
Shearling Ewes home-bred	12	Sir Model (939)
Total	27	
Lambs bred kept as Rams	13	
Ewe Lambs	20	

Messrs. McGREGOR BROS.,

NGUTUAWA, MASTERTON.

Flock No. 100.

This flock was founded in 1904 by the purchase of 30 ewe and 15 ewe lambs bred by Mr. Isaac Sykes (Flock No. 48).

1906—Ewes retained from previous year	35	**Sire—**
Shearling Ewes home-bred	12	Nero (679)
Total	47	
Lambs bred kept as Rams	23	
Ewe Lambs	26	

Messrs. McKENZIE & LOVELOOK,

CLIFF SIDE, PALMERSTON NORTH.

Flock No. 101.

This flock was founded in 1905 by the purchase of 103 ewes bred by Mr. D. Rowland (Flock No. 50), and 50 ewes bred by Mr. W. Akers (Flock No. 73), and 18 ewe hoggets and 20 ewes with 32 lambs bred by Mr. J. O. Batchelar (Flock No. 45). In 1907, 100 2-tooth ewes were purchased from Mr. R. Tanner (Flock No. 63).

			Sires—
1906—Ewes retained from previous year	173		
Ewes purchased		18	Willowbank 1 (987)
	Total	191	Willowbank 2 (988)
Lambs bred kept as Rams	111		Willowbank 3 (989)
Ewe Lambs	125		Tiritea Duke 2nd (454)
			Bealey No. 1 (1315)

Messrs. McKINNON BROS.,

OHINEWAI, WAIKATO.

Flock No. 102.

This flock was founded in 1903 by the purchase of 13 ewes bred by the New Zealand Lands Association, Limited (Flock No. 80).

		Sire—
1906—Ewes retained from previous year	15	
Ewes put to Rams	15	The Chief C 28 (927)
Lambs bred kept as Rams	8	
Ewe Lambs	11	

Mr. JOSEPH MITCHELL,

BROCKHURST, MARTON.

Flock No. 103.

This flock was founded in 1905 by the purchase of 49 ewes from Mr. Wm. F. Jacob (Flock No. 43).

			Sires—
1906—Ewes retained from previous year		48	
Ewes put to Rams	-	48	Bondsman (147)
Lambs bred kept as Rams	30		No. 2319 (169)
Ewe Lambs	32		

Mr. WALTER G. PEARCE,

CLOVERLY, COLYTON.

Flock No. 104.

This flock was founded in 1904 by the purchase of 30 ewes in lamb at the dispersion sale of the Woodendean Flock (No. 26).

In 1906, 7 ewe lambs were purchased from Mr. D. J. Willis (Flock No. 117).

		Sire—
1906—Ewes retained from previous year	30	
Shearling Ewes home-bred	15	The Nut (1317)
Total	45	
Lambs bred kept as Rams	18	
Ewe Lambs	23	

Mr. WILLIAM H. RAYNER,

TARATAHI, MASTERTON.

Flock No. 105.

This flock was founded in 1904 by the purchase of 10 two-tooth ewes bred by Mr. G. Wheeler (Flock No. 24).

		Sire—
1906—Ewes retained from previous year	10	
Shearling Ewes home-bred	5	Parorangi 8th (644)
Total	15	
Lambs bred kept as Rams	9	
Ewe Lambs	8	

Mr. WILLIAM REID,

MAKINO.

Flock No. 106.

This flock was founded in 1904 by the purchase of 20 ewes in lamb at the dispersion sale of the Woodendean Flock (No. 26).

In 1905, 22 ewes, in lamb, from the same flock, were purchased from the Executors of the late Mr. W. Baker.

1906—Ewes retained from previous year		39	Sire—
Ewes put to Rams		39	Waituna (643)
Lambs bred kept as Rams	20		
Ewe Lambs	24		

Mr. FREDERICK RICHARDSON,

WAITUNA WEST.

Flock No. 107.

This flock was founded in 1904 by the purchase of 26 ewes in lamb at the dispersion sale of the Woodendean Flock (No. 26).

1906—Ewes retained from previous year		23	Sire—
Shearling Ewes home-bred		9	Musketeer (153)
Total		32	
Lambs bred kept as Rams	17		
Ewe Lambs	22		

Mr. ERNEST SHORT,

BALMORAL, PAKIHIKURA.

Flock No. 109.

The Balmoral Flock was founded in 1904 by the purchase of 10 ewes in lamb, 30 maiden shearling ewes, and 23 ewe lambs, at the dispersion sale of the Woodendean Flock (No. 26).

1906—Ewes retained from previous year		55	Sire—
Shearling Ewes home-bred		7	The Duke (935)
Total		62	
Lambs bred kept as Rams	38		
Ewe Lambs	22		

Mrs. B. H. SLACK,

TAIKOREA, OROUA BRIDGE.

Flock No. 110.

This flock was founded in 1904 by the purchase of 50 ewes bred by Mr. Robt. Cobb (Flock No. 22). In 1905, 50 ewes, and in 1906, 2 ewes from the same flock were added. In 1905, 7 ewes were purchased from the Executors of the late Mr. W. Baker (Flock No. 88), also 3 ewes from Mr. J. O. Batchelar (Flock No. 45).

			Sires—
1906—Ewes retained from previous year		90	King Cobb No. 110 (992)
Shearling Ewes home-bred		31	Model (662)
	Total	121	
30 of above sold in lamb in August		30	
		91	
Lambs bred kept as Rams	56		
Ewe Lambs	40		

Messrs. SMITH BROS.,

WOODLANDS, COLYTON.

Flock No. 111.

This flock was founded in 1905 by the purchase of 23 ewe lambs from Mr. D. G. Short (Flock No. 108). In 1906, 7 ewes, bred by Mrs. Willis (Flock No. 26), were purchased from Mr. Jas. Bell (Flock No. 89).

			Sire—
1906—Ewes purchased		30	J.O.B. 338 G (1329)
Ewes put to Rams		30	
Lambs bred kept as Rams	11		
Ewe Lambs	16		

Mr. T. ALFRED SMITH,

COLYTON.

Flock No. 112.

This flock was founded in 1905 by the purchase of 6 ewes and 5 ewe lambs from Mr. D. G. Short (Flock No. 108). In 1906, 25 ewes purchased from Mr. G. Wheeler (Flock No. 24), and 10 ewes from Mr. Jas. Bell (Flock No. 89), were added.

			Sire—
1906—Ewes retained from previous year		6	
Shearling Ewes home-bred		5	Willis 1-3 (993)
Ewes purchased		35	
	Total	46	
Lambs bred kept as Rams	12		
Ewe Lambs	6		

Mr. ALFRED TYER,

FEATHERSTON.

Flock No. 114.

This flock was founded in 1905 by the purchase of 10 four-tooth ewes from Mr. G. Wheeler (Flock No. 24).

			Sire—
1906—Ewes purchased		10	
Ewes put to Rams		10	Westminster (564)
Lambs bred kept as Rams	4		
Ewe Lambs	6		

Mr. MAX VOSS,

KARERE, LONGBURN.

Flock No. 115.

The Karere flock was founded in 1905 by the purchase of 30 two-tooth ewes, bred by Mr. D. Rowland (Flock No. 50), and a previous purchase of 50 ewes from the same flock, and was accepted by the Council the same year.

			Sires—
1906—Ewes retained from previous year		68	
Shearling Ewes home-bred		27	Elderslie 2nd No. 1905 (1333)
	Total	95	J.O.B. 369 G (1000)
Lambs bred kept as Rams	58		
Ewe Lambs	77		

Mr. HENRY O. WILKINSON,

HINABURN, FEATHERSTON.

Flock No. 116.

This flock was founded in 1905 by the purchase of 70 ewes bred by Mr. Jas. Wilkinson (Flock No. 57).

		Sires—
1906—Ewes retained from previous year	65	Lakewood 1st (1001)
Ewes put to Rams	65	Lakewood 2nd (1002)
Ewe Lambs	45	

Mr. DANIEL J. WILLIS,

HAWERA.

Flock No. 117.

This flock is now dispersed.

Mr. THOS. R. WILLIS,

WOODENDEAN, GREATFORD.

Flock No. 118.

This flock was founded in 1904 by the purchase of 37 ewes at the dispersion sale of the Woodendean Flock (No. 26).

		Sires—
1906—Ewes retained from previous year	29	Willis 183 (990)
Shearling Ewes home-bred	13	Primate 6th (554)
Total	42	
Lambs bred kept as Rams	28	
Ewe Lambs	18	

Mr. JAMES G. WILSON,

BURLEIGH, BULLS.

Flock No. 119.

The Burleigh flock was founded in 1904 by the purchase of 10 two-tooth ewes from Mr. G. Wheeler (Flock No. 24), and 19 ewes in lamb, also 19 ewe lambs at the dispersion sale of the Wood-endean Flock (No. 26).

In 1905, 19 ewes in lamb, bred by Mr. G. Wheeler (Flock No. 24) were added.

		Sires—
1906—Ewes retained from previous year	61	
Shearling Ewes home-bred	12	Primate VI. (554)
	—	Willowbank (1334)
Total	73	Old Constitution (1335)
Lambs bred kept as Rams	30	
Ewe Lambs	38	

Mr. WILLIAM WINDLEY,

PORIRUA.

Flock No. 120.

This flock was founded in 1905 by the purchase of 17 ewes from Mr. T. P. Allen (Flock No. 35).

		Sire—
1906—Ewes retained from previous year	17	
Ewes put to Rams	17	Woolpack (1003)
Lambs bred kept as Rams	7	
Ewe Lambs	13	

Mr. ALFRED E. HARDING,

MANGAWHARE, KAIPARA, AUCKLAND.

Flock No. 121.

The Mangawhare flock was founded in 1889 by the purchase of 50 stud ewes from the late Mr. John Harding, Mount Vernon (Flock No. 72), and a further purchase in 1903 of 33 ewe lambs at the dispersion sale of the Woodlands Flock (No. 80).

Previous to entry with this Association, rams were obtained from Messrs. J. Harding, F. Hutchinson, Mungavin, Gray Bros., and Bealey Bros.

1906—Ewes retained from previous year	140	Sires—
Shearling Ewes home-bred	60	Porirua (1339)
		Aoroa 2nd (1338)
Total	200	Aoroa 1st (700)
Lambs bred kept as Rams	106	Major 5th (634)
Ewe Lambs	81	Major (323)
		Favorite 2nd (635)

Mr. THOMAS HUNT,

HIGHFIELD, WAKEFIELD, NELSON.

Flock No. 122.

This flock of 224 ewes, entered in N.Z. Flock Book, South Island Volume, 1904/5, Flock No. 100, was founded about 1875 by the purchase from Mr. Roderick McRae, of Richmond, of ewes descended from Dr. Webster's flock. In 1890 more ewes of the same strain were added.

Rams used previous to entry with this Association were bred by Messrs. R. McRae, A. Matthews, E. W. Dorset, Gray Bros., H. Eglinton, G. Wheeler, Rowland, Bealey Bros., Tanner, Reid, and Mrs. Willis.

1906—Ewes retained from previous year	152	Sires—
Shearling Ewes home-bred	72	Champion (642)
		Lord Nelson 986 (942)
Total	224	E.W.D. 04 (1004)
Lambs bred kept as Rams	111	Batchelar 1 (1005)
Ewe Lambs	119	

Mr. JOSIAH BATCHELAR,

BROOKLEIGH, LINTON.

Flock No 123.

This flock was founded in 1905 by the purchase of 11 ewes, bred by Mr. J. O. Batchelar, at the dispersion sale of the Tiritea Flock (No. 45).

1906—Ewes retained from previous year	11	Sire—
Ewes put to Rams	11	Prince Imperial (176)
Lambs bred kept as Rams	5	
Ewe Lambs	5	

Messrs. DUNCAN & CAMPION,

OKIRAE, FORDELL.

Flock No. 124.

This flock was founded in 1905 by the purchase of 41 ewes and their progeny from Mr. P. C. Neill, who obtained them at the dispersion sale of the Woodendean Flock (No. 26) in 1904.

1906—Ewes retained from previous year	40	Sire—
Shearling Ewes home-bred	14	Comrade No. 47 (1341)
Total	54	
Lambs bred kept as Rams	31	
Ewe Lambs	25	

Mr. W. A. ELLIS,

YORK FARM, MARTON.

Flock No. 125.

This flock was founded in 1905 by the purchase of 16 ewes with lambs at the dispersion sale of Mr. J. O. Batchelar (Flock No. 45), and in January, 1906, 100 ewes bred by Mr. D. Rowland (Flock No. 50) were added.

1906—Ewes retained from previous year	16	Sires—
Ewes purchased	100	Kuroki (602)
Total	116	Primate 5th (553)
Lambs bred kept as Rams	64	Burford (530)
Ewe Lambs	79	Loseley (1342)

Mr. WILLIAM GIBSON,

KIWITEA.

Flock No. 126.

This flock was founded in 1905 by the purchase of 50 ewes with 14 ewe lambs, bred by Mr. J. O. Batchelar, at the dispersion of the Tiritea Flock (No. 45).

1906—Ewes purchased		50	Sire—
Ewes put to Rams		50	Beck Dyke (1343)
Lambs bred kept as Rams	40		
Ewe Lambs	32		

Mr. JAMES CHARLES KELLY,

TOKOMARU.

Flock No. 127.

This flock was founded in 1905 by the purchase from the Executors of the late Mr. W. Baker (Flock No. 88) of 24 ewes bred by Mrs. Willis (Flock No. 26), and also of 20 ewes, bred by Mr. J. O. Batchelar, at the dispersion sale of the Tiritea Flock (No. 45).

In 1906, 24 ewes were purchased from Mr. Robt. Tanner (Flock No. 63).

1906—Ewes retained from previous year		44	Sire—
Ewes purchased		24	Tokomaru (1349)
	Total	68	
Lambs bred kept as Rams	32		
Ewe Lambs	33		

Mr. FREDERICK W. H. KUMMER,

MAURICEVILLE.

Flock No. 128.

This flock was founded in 1905 by the purchase of 10 ewes with lambs, bred by Mr. J. O. Batchelar, at the dispersion sale of the Tiritea Flock (No. 45).

1906—Ewes retained from previous year		10	Sire—
Ewes put to Rams		10	Mahoe I. (1350)
Lambs bred kept as Rams	11		
Ewe Lambs	4		

Mr. CHARLES LUCAS,
PAKIHIKURA.

Flock No. 129.

This flock is now dispersed.

Messrs. McLEAN BROS.,
WAITUNA WEST.

Flock No. 130.

This flock was founded in 1905 by the purchase of 25 ewes, bred by Mr. J. O. Batchelar, at the dispersion sale of the Tiritea Flock (No. 45).

1906—Ewes retained from previous year	25	**Sire—**
Ewes put to Rams	25	
Lambs bred kept as Rams	14	Windsor A 46 (914)
Ewe Lambs	16	

Mr. MAURICE MASON
TAHEKE, PUKEHOU, HAWKE'S BAY.

Flock No. 131.

This flock was founded in 1905 by the purchase of 6 ewe hoggets, bred by Mr. J. O. Batchelar, at the dispersion sale of the Tiritea Flock (No. 45).

1906—Ewes purchased	6	**Sire—**
Ewes put to Rams	6	
Lambs bred kept as Rams	1	G 307 (1351)
Ewe Lambs	4	

Mr. OWEN M. MONCKTON,

PATUTAHI, GISBORNE.

Flock No. 132.

This flock was founded in 1902 with 57 ewe hoggets, bred by Mr. F. Monckton, and having been passed in at the dispersion sale of his Flock (No. 32), were transferred to present owner.

			Sires—
1905—Ewes retained from previous year		62	
Shearling Ewes home-bred		20	Ram bred by R. Cobb (Flock 22)
		—	Fairfield Ram
	Total	82	
Lambs bred kept as Rams	32		
Ewe Lambs	46		
1906—Ewes retained from previous year		80	Rissington (1352)
Shearling Ewes home-bred		32	Fairfield (1353)
	Total	112	
Lambs bred kept as Rams	46		
Ewe Lambs	58		

Mr. J. Q. OATES,

PEACH GROVE, CARTERTON.

Flock No. 133.

This flock was founded in 1906 by the purchase of 5 ewes from Mr. W. B. Allen (Flock No. 36).

			Sire—
1906—Ewes purchased		5	
Ewes put to Rams		5	E.W.D. 026 (1354)
Lambs bred kept as Rams	1		
Ewe Lambs	5		

Mr. W. B. V. PEAROE,
OROUA BRIDGE.

Flock No. 134.

This flock was founded in 1905 by the purchase of 114 ewes and also 20 ewe hoggets, bred by Mr. J. O. Batchelar, at the dispersion sale of the Tiritea Flock (No. 45).

In 1906, 1 ewe and 1 ewe lamb were purchased from Mr. Cobb (Flock No. 22).

1906—Ewes retained from previous year	114	**Sire—**
Shearling Ewes home-bred	20	Utiku (1355)
Ewes purchased	1	
	Total 135	
Lambs bred kept as Rams	50	
Ewe Lambs	60	

Mr. JAMES LEWIS PERRY,
WAIMIRO, WAIMATA VALLEY, GISBORNE.

Flock No. 135.

This flock was founded in 1906 by the purchase of 100 ewes from Mr. G. C. Wheeler (Flock No. 24).

1906—Ewes purchased	100	**Sires—**
Ewes put to Rams	100	Wantage (561)
Lambs bred kept as Rams	46	Godolphin (419)
Ewe Lambs	23	

Mr. GEO. N. REYNOLDS,
SANDOWN, MANGAHEIA, GISBORNE.

Flock No. 136.

This flock was founded in 1905 by the purchase of 86 ewes with lambs, bred by Mr. J. O. Batchelar, at the dispersion sale of the Tiritea Flock (No. 45).

1906—Ewes purchased	85	**Sires—**
Ewes put to Rams	85	Burford No. 441 (817)
Lambs bred kept as Rams	37	Burford No. 462 (821)
Ewe Lambs	42	

Messrs. SMITH BROS.,

PAIKAKARIKI.

Flock No. 137.

The Wainui flock was founded in 1906 by the purchase of 54 ewes from Mr. G. Wheeler (Flock No. 24), and 22 ewes from Mr. H. S. Hadfield (Flock No. 49).

1906—Ewes purchased		76	Sire—
Ewes put to Rams		76	
Lambs bred kept as Rams	45		Tostig (605)
Ewe Lambs	32		

Mr. GERALD TOLHURST,

RAKANA, GLEN OROUA.

Flock No. 138.

This flock was founded in 1905 by the purchase of 10 ewes from Messrs. McHardy Bros. (Flock No. 79).

1906—Ewes retained from previous year		10	Sires—
Ewes put to Rams		10	
Lambs bred kept as Rams	5		Rakana 1 (1356)
Ewe Lambs	7		Rakana 2 (1357)

Mr. G. A. WHEELER,

RANGITANE, KAWHATAU, MANGAWEKA.

Flock No. 139.

This flock was founded in 1905 by the purchase of 5 ewe hoggets, bred by Mr. D. J. Willis (Flock No. 117), and 5 ewes from the Burford Flock (No. 24).

Ewes purchased		10	Sires—
Ewes put to Rams		10	
Lambs bred kept as Rams	1		Imperialist (794)
Ewe Lambs	5		Freeholder (797)

Mr. W. G. AITKEN,

ASHHURST.

Flock No. 140.

This flock was founded in 1906 by the purchase of 12 ewes and 30 ewe lambs from Mr. James Bell (Flock No. 89).

1906—Ewes purchased		12	**Sire—**
Ewes put to Rams		12	Fairburn ram (1358)
Lambs bred kept as Rams	6		
Ewe Lambs	8		

Mr. JOHN ALLEN,

"THE CLIFFS," WAINGARO, WAIKATO.

Flock No. 141.

This flock was founded in 1907 by the purchase, from Messrs. McKenzie and Lovelock, of 40 ewes, bred by Mr. D. Rowland (Flock No. 50), 28 ewes were also purchased from Mr. J. Kebbell (Flock No. 40).

Mr. G. E. ALLEN,

TIPUA, CLAREVILLE, WAIRARAPA.

Flock No. 142.

This flock was founded in 1907 by the transfer of 208 ewes and 89 ewe lambs from Mr. W. B. Allen (Flock No. 36).

Mr. ERNEST WILLIAM ALLEN,

DEALWOOD, CLAREVILLE, WAIRARAPA.

Flock No. 143.

This flock was founded in 1907 by the transfer of 43 ewes and 25 ewe lambs from Mr. W. B. Allen (Flock No. 36).

Mr. E. RUI BATLEY,

MOAWHANGO.

Flock No. 144.

This flock was founded in 1906 by the purchase of 20 ewes and 10 ewe lambs from Mr. E. Short (Flock No. 77).

1906—No returns.

Mr. W. E. BAKER,

MAKINO.

Flock No. 145.

This flock was founded in 1907 by the purchase of 7 ewes from Mr. Wm. F. Jacob (Flock No. 43).

Mr. W. E. BIDWELL,

ROTOTAWAI, FEATHERSTON.

Flock No. 146.

This flock was founded in 1895 and is descended from the flock of the late Mr. C. R. Bidwell.

1906—Ewes retained from previous year	14		**Sire—**
Shearling Ewes home-bred	5		Parorangi No. 897 (1359)
	Total	19	
Lambs bred kept as Rams	12		
Ewe Lambs	9		

Mr. W. H. BOOTH,

MIDDLE RUN, CARTERTON.

Flock No. 147.

This flock was founded in 1906 by the purchase of 5 ewes from Mr. W. J. Nix (Flock No. 39), and of 5 ewes from Mr. W. C. Dorset (Flock No. 25).

1906—Ewes purchased		10	**Sire—**
Ewes put to Rams		10	
Lambs bred kept as Rams	3		No. 1 Gray (1360)
Ewe Lambs	3		

Mr. W. H. BUIOK,

MASTERTON.

Flock No. 148.

This flock was founded in 1906 by the purchase, from Mrs. B. H. Slack, of 22 ewes bred by Mr. R. Cobb (Flock No. 22), 7 ewes bred by the late Mr. W. Baker (Flock No. 88), and 1 ewe bred by Mrs. B. H. Slack (Flock No. 110), also by the purchase from Mr. W. B. Allen (Flock No. 36) of 14 ewes.

1906—Ewes purchased		44	**Sire—**
Ewes put to Rams		44	
Lambs bred kept as Rams	22		Trueboy (340)
Ewe Lambs	32		

Mr. E. A. CAMPBELL,

WIRITOA, WANGANUI.

Flock No. 149.

This flock was founded in 1906 by the purchase of 20 ewes from Mr. James Bell (Flock No. 89).

1906—Ewes purchased		20	Sire—
Ewes put to Rams		20	Brandon Hall No. 87 (1361)
Lambs bred kept as Rams	6		
Ewe Lambs	5		

Mr. G. L. COOK,

WHAKAPUNI, HUNTERVILLE.

Flock No. 150.

This flock was founded in 1906 by the purchase of 22 ewe hoggets from Mr. J. Mitchell (Flock No. 103).

Mr. EDWARD CRESWELL,

FOXTON.

Flock No. 151.

This flock was founded in 1907 by the purchase of 3 ewe lambs from Mr. W. B. V. Pearce (Flock No. 134).

Mr. O. G. O. DERMER,

CLOVERDENE, FEILDING.

Flock No. 152.

This flock was founded in 1907 by the purchase of 19 ewes from Mr. D. P. Buchanan (Flock No. 91).

Mr. E. EAGLE, Junr.,

BELVEDERE, CARTERTON.

——

Flock No. 153.

This flock was founded in 1906 by the purchase of 10 ewes from Mr. W. J. Nix (Flock No. 39).

1906—Ewes purchased		10	**Sire—**
Ewes put to Rams		10	Belvedere (1362)
Lambs bred kept as Rams	5		
Ewe Lambs	6		

Mr. H. B. EGLINTON,

KAITAWA, PAHIATUA.

——

Flock No. 154.

This flock was founded in 1906 by the purchase of 30 ewes from Mr. A. Matthews (Flock No. 21).

1906—Ewes purchased		30	**Sire—**
Ewes put to Rams		30	Fairburn (206)
Lambs bred kept as Rams	23		
Ewe Lambs	20		

Mr. H. R. ELDER,

WAIMAHOE, WAIKANAE.

——

Flock No. 155.

This flock was founded in 1907 by the purchase of 30 ewes from Mr. G. Wheeler (Flock No. 24), and of 20 ewes from Mr D. Rowland (Flock No. 50).

Mr. D. H. GARDNER,

PAIAKA, KOPUTAROA.

Flock No. 156.

This flock was founded in 1907 by the purchase of 10 ewes from Mr. D. Rowland (Flock No. 50).

Mr. A. HARDING,

SIBERIA, ASHHURST.

Flock No. 157.

This flock was founded in 1906 by the purchase of 20 ewe hoggets from Messrs. Gray Bros. (Flock No. 38).

Mr. D. H. KILGOUR,

KIWITEA.

Flock No. 158.

This flock was founded in 1907 by the purchase of 25 ewes from Messrs. McHardy Bros. (Flock No. 79).

Mr. O. P. LYNCH,

EMERALD GLEN, PAIKAKARIKI.

Flock No. 159.

This flock was founded in 1906 by the purchase of 15 ewes and 10 ewe lambs from Mr. T. P. Allen (Flock No. 35.)

1906—Ewes purchased		15	Sire—
Ewes put to Rams		15	Waimahaka No. 213 (624)
Lambs bred kept as Rams	9		
Ewe Lambs	9		

Mr. LESLIE McHARDY,

BLACKHEAD, WAIPAWA.

Flock No. 160.

This flock was founded by Messrs. McHardy Bros. at Palmerston North in 1901, by the purchase of 120 ewes and 36 ewe lambs from Mr. Wakelin when Flock No. 59 was dispersed.

In 1906, 9 ewes were purchased from Mr. Cobb (Flock No. 22).

Previous to entry with this Association, rams bred by Messrs. Dorset and Sykes have been used in this flock.

In 1907 half of this flock was transferred by Mr. Leslie McHardy to the above address.

Mr. W. J. A. McGREGOR,

MOUNT LINTON, SOUTHLAND.

Flock No. 161.

This flock was founded in 1907 by the purchase of 40 ewes from Mr. J. Holms (Flock No. 1).

Mr. W. E. SMITH,

BRUNSWICK, WANGANUI.

Flock No. 162.

This flock was founded in 1907 by the purchase of 27 ewes from Mr. G. Wheeler (Flock No. 24).

Mr. S. STANDEN,

RUAPUHA, FEILDING.

Flock No. 163.

This flock was founded in 1907 by the purchase of 11 ewes from Messrs. McKenzie & Lovelock, bred by Mr. D. Rowland (Flock No. 50), and 35 ewes from Mr. R. Tanner (Flock No. 63).

Messrs. J. STUCKEY & SON,

OPAKI, WAIRARAPA.

Flock No. 164.

This flock was founded in 1906 by the purchase of 20 ewes from Mr. G. Wheeler (Flock No. 24).

1906—Ewes purchased		20	Sires—
Ewes put to Rams		20	
Lambs bred kept as Rams	13		Watchman 2nd (828)
Ewe Lambs	17		Landlord (810)
			Burford No. 227 (1163)

This flock is now dispersed.

Mr. T. WAUGH,

KIMBOLTON.

Flock No. 165.

This flock was founded in 1906 by the purchase of 5 ewes from Mr. D. J. Willis (Flock No. 117).

1906—Ewes purchased		5	Sire—
Ewes put to Rams		4	Burford 1618 (86)
Lambs bred kept as Rams	2		
Ewe Lambs	2		

Mr. JAMES F. WHITE,

OHAURA, GREYMOUTH, WESTLAND.

Flock No. 166.

This flock was founded in 1907 by the purchase of 10 ewes from Mr. Thos. Hunt (Flock No. 122).

Mr. J. O. BATCHELAR,

TIRITEA, PALMERSTON NORTH.

Flock No. 167.

This flock was started in 1907 with 8 ewes left in Mr. Batchelar's hands after his dispersion sale in 1905.

1906—No increase.

SHEEP FOR SEPARATE ENTRY.

———o———

DIRECTIONS FOR THE READER.

Numbers enclosed in parentheses (   ) refer to the Flock Book.   Numbers without parentheses refer to Private Registers.

———o———

ROMNEY MARSH.

———

RAMS.

———

Mr. JAMES HOLMS.   Flock. No. 1.

———

Flock
Book No.

Endurance (No. 819), born 1904.

Bred by Mr. G. Wheeler.

Sire, Waterloo

Dam, a Burford ewe

Mr. JOHN F. REID.   Flock No. 5.

———

Flock
Book No.

1041   No. 255, born 1904.

1042   No. 256, born 1904.

1043   No. 263, born 1904.

1044   No. 271, born 1904.

1045   No. 264, born 1904.

1046   No. 257, born 1904.

1047   No. 265, born 1904.

1048   No. 270, born 1904.

———

Mrs. JAMES FULTON,   Flock No. 6.

(Catherine Fulton, Widow of Hon. James Fulton.)

Mr. H. V. Fulton, Manager.

———

1049   Ajax III., born 1904.

Sire, King Dick (1037)

Dam, a Raviffenscle ewe

Messrs. BEALEY BROS. Flock No. 9.

Flock
Book No.

1050   **Farmer's 26 of 1903 (imp.).**

Bred by Mr. G. Farmer, Leeds Abbey, Maidstone, Kent.

Sire, Macknade's 21st

Dam, a Leeds Abbey ewe

1051   **Westbroke 86 of 1905 (imp.).**

Bred by Mr. Arthur Finn, Westbroke, Lydd, Kent.

Sire, Westbroke 85 of 1901

Dam, Windsor 1st (6899)

Mr. J. R. MACKENZIE. Flock No. 19.

1052   **Elham No. 84 of 1906.**

Bred by Mr. Charles File, Elham, Kent, England.

Sire, Elham No. 58 (15008), Vol. 11

Dam, Elham ewe (1060)

1053   **Elham No. 86 of 1906.**

Bred by Mr. Charles File, Elham, Kent, England.

Sire, Elham No. 58 (15008), Vol. 11

Dam, Elham ewe 11⁰,pp 22, Vol. 11

1054   **Lyndale No. 110 of 1905.**

Bred by Mr. W. Millen, Lyndale, Kent.

Sire, Lyndale No. 13 of 1903

Dam, a Millen ewe, pp 25, Vol. 10

1055   **Lyndale No. 55 of 1905.**

Bred by Mr. W. Millen, Lyndale, Kent.

Sire, Macknade Bean 12th (12684), Vol. 9

Dam, a Millen ewe 100^2 pp 25, Vol. 10

1056   **Ospringe No. 307 of 1905.**

Bred by Mr. J. P. Barnes, Ospringe, Kent.

Macknade Blue Knob 4th (13611), Vol. 10

Dam, a Barnes ewe (113^2) pp 26, Vol. 11

1057   **Garrington No. 36 of 1905.**

Bred by Mr. J. D. Maxted, Lower Garrington, Kent.

Sire, Amos No. 59 of 1903 (13849), Vol. 10

Dam, a Garrington ewe, pp 21, Vol. 11

Mr. A. MATTHEWS.   Flock No. 21.

1058   **No. 905, born 1905.**

Sire, Westbroke No. 1

Dam, a Waiorongomai ewe

1059   **No. 477, born 1905.**

Sire, Windsor 83rd

Dam, a Waiorongomai ewe

1060   **No. 563, born 1905.**

Sire, No. 224 of 1900

Dam, a Waiorongomai ewe

1061   **No. 516, born 1905.**

Sire, Windsor 83rd

Dam, a Wairorongomai ewe

Flock
Book No.

1062   **No. 872,** born 1905.

Sire, Hazlewood 42nd

Dam, a Waiorongomai ewe

1063   **No. 717,** born 1905.

Sire, No. 96 of 1900

Dam, a Waiorongomai ewe

1064   **No. 614,** born 1905.

Sire, No. 224 of 1900

Dam, a Waiorongomai ewe

1065   **No. 572,** born 1905.

Sire, No. 5 of 1900

Dam, a Waiorongomai ewe

1066   **No. 884,** born 1905.

Sire, Westbroke No. 1

Dam, a Waiorongomai ewe

1067   **No. 859,** born 1905.

Sire, Westbroke No. 9

Dam, a Waiorongomai ewe

1068   **No. 666,** born 1905.

Sire, No. 142 of 1900

Dam, a Waiorongomai ewe

1069   **No. 558,** born 1905.

Sire, No. 224 of 1900

Dam, a Waiorongomai ewe

1070   **No. 557,** born 1905.

Sire, No. 5 of 1900

Dam, a Waiorongomai ewe

Flock
Book No.

1071   **No. 443,** born 1905.
                                   Sire, Westbroke No. 9
     Dam, a Waiorongomai ewe

1072   **No. 613,** born 1905,
                                   Sire, No. 224 of 1900
     Dam, a Waiorongomai ewe

1073   **No. 610,** born 1905.
                                   Sire, No. 224 of 1900
     Dam, a Waiorongomai ewe

1074   **No. 519,** born 1905.
                                   Sire, Windsor 83rd
     Dam, a Waiorongomai ewe

1075   **No. 904,** born 1905.
                                   Sire, Westbroke No. 1
     Dam, a Waiorongomai ewe

1076   **No. 882,** born 1905.
                                   Sire, Westbroke No. 1
     Dam, a Waiorongomai ewe

1077   **No. 908,** born 1905.
                                   Sire, Westbroke No. 1
     Dam, a Waiorongomai ewe

1078   **No. 755,** born 1905.
                                   Sire, No. 227 of 1903
     Dam, a Waiorongomai ewe

1079   **No. 800,** born 1905.
                                   Sire, Hazlewood 44th
     Dam, a Waiorongomai ewe

Flock
Book No.

1080   **No. 428,** born 1905.

Sire, Westbroke No. 9

Dam, a Waiorongomai ewe

1081   **No. 932,** born 1905.

Sire, Westbroke No. 1

Dam, a Waiorongomai ewe

1082   **No. 175,** born 1904.

Sire, Westbroke No. 1

Dam, a Waiorongomai ewe

1083   **No. 54,** born 1904.

Sire, Westbroke No. 9

Dam, a Waiorongomai ewe

1084   **No. 172,** born 1904.

Sire, Westbroke No. 1

Dam, a Waiorongomai ewe

1085   **No. 235,** born 1904.

Sire, Westbroke No. 9

Dam, a Waiorongomai ewe

1086   **No. 189,** born 1904.

Sire, Westbroke No. 1

Dam, a Waiorongomai ewe

1087   **No. 148,** born 1904.

Sire, Nackington No. 5

Dam, a Waiorongomai ewe

1088   **No. 125,** born 1904.

Sire, Nackington No. 5

Dam, a Waiorongomai ewe

1089   **No. 565,** born 1905.
                                        Sire, No. 224 of 1900
          Dam, a Waiorongomai ewe

1090   **No. 911,** born 1905.
                                        Sire, Westbroke No. 1
          Dam, a Waiorongomai ewe

1091   .   **No. 790,** born 1905.
                                        Sire, No. 227 of 1903
          Dam, a Waiorongomai ewe

1092   **No. 845,** born 1905.
                                        Sire, Hazlewood 44th
          Dam, a Waiorongomai ewe

1093   **No. 629,** born 1905.
                                        Sire, No. 224 of 1900
          Dam, a Waiorongomai ewe

1094   **No. 758,** born 1905.
                                        Sire, No. 227 of 1903
          Dam, a Waiorongomai ewe

1095   **No. 640,** born 1905.
                                        Sire, No. 142 of 1900
          Dam, a Waiorongomai ewe

1096   **No. 695,** born 1905.
                                        Sire, No. 96 of 1900
          Dam, a Waiorongomai ewe

1097   **No. 822,** born 1905.
                                        Sire, Hazlewood 44th
          Dam, a Waiorongomai ewe

Flock
Book No.

1098   **No. 835,** born 1905.
                                             Sire, Hazlewood 44th
        Dam, a Waiorongomai ewe

1099   **No. 551,** born 1905.
                                             Sire, No. 5 of 1900
        Dam, a Waiorongomai ewe

1100   **No. 906,** born 1905.
                                             Sire, Westbroke No. 1
        Dam, a Waiorongomai ewe

1101   **No. 839,** born 1905.
                                             Sire, Hazlewood 44th
        Dam, a Waiorongomai ewe

1102   **No. 699,** born 1905.
                                             Sire, No. 96 of 1900
        Dam, a Waiorongomai ewe

1103   **No. 892,** born 1905.
                                             Sire, Westbroke No. 1
        Dam, a Waiorongomai ewe

1104   **No. 407,** born 1905.
                                             Sire, Westbroke No. 9
        Dam, a Waiorongomai ewe

1105   **No. 642,** born 1905.
                                             Sire, No. 142 of 1900
        Dam, a Waiorongomai ewe

1106   **No. 455,** born 1905.
                                             Sire, Rigden's No. 83
        Dam, a Waiorongomai ewe

1107   **No. 861,** born 1905.
                                        Sire, Westbroke No. 9
       Dam, a Waiorongomai ewe

1108   **No. 506,** born 1905.
                                        Sire, Windsor No. 83
       Dam, a Waiorongomai ewe

1109   **No. 840,** born 1905.
                                        Sire, Hazlewood 44th
       Dam, a Waiorongomai ewe

1110   **No. 454,** born 1905.
                                        Sire, Rigden's No. 83
       Dam, a Waiorongomai ewe

1111   **No. 427,** born 1905.
                                        Sire, Westbroke No. 9
       Dam, a Waiorongomai ewe

1112   **No. 421,** born 1905.
                                        Sire, Westbroke No. 9
       Dam, a Waiorongomai ewe

1113   **No. 554,** born 1905.
                                        Sire, No. 224 of 1900
       Dam, a Waiorongomai ewe

1114   **No. 781,** born 1905.
                                        Sire, No. 227 of 1903
       Dam, a Waiorongomai ewe

1115   **No. 673,** born 1905.
                                        Sire, No. 142 of 1900
       Dam, a Waiorongomai ewe

1116   **No. 667,** born 1905.

                           Sire, No. 142 of 1900

Dam, a Waiorongomai ewe

1117   **No. 805,** born 1905.

                           Sire, Hazlewood 44th

Dam, a Waiorongomai ewe

1118   **No. 633,** born 1905.

                           Sire, No. 142 of 1900

Dam, a Waiorongomai ewe

1119   **No. 401,** born 1905.

                           Sire, No. 227 of 1903

Dam, a Waiorongomai ewe

1120   **No. 425,** born 1905.

                           Sire, Westbroke No. 9

Dam, a Waiorongomai ewe

1121   **No. 843,** born 1905.

                           Sire, Hazlewood 44th

Dam, a Waiorongomai ewe

1122   **No. 760,** born 1905.

                           Sire, No. 227 of 1903

Dam, a Waiorongomai ewe

1123   **No. 403,** born 1905.

                           Sire, Westbroke No. 9

Dam, a Waiorongomai ewe

1124   **No. 652,** born 1905.

                           Sire, No. 142 of 1900

Dam, a Waiorongomai ewe

1125   **No. 767,** born 1905.

                           Sire, No. 227 of 1903

Dam, a Waiorongomai ewe

Mr. ROBT. COBB.   Flock No. 22.

Flock
Book No.

1126   **Buffalo Bill VI., born 1905.**
                                   Sire, Buffalo Bill V.
       Dam, a Kent Stud Farm ewe

1127   **Active V., born 1905.**
                                   Sire, Active IV.
       Dam, a Kent Stud Farm ewe

1128   **Jumbo XI., born 1905.**
                                   Sire, Jùmbo IX.
       Dam, a Kent Stud Farm ewe

EXECUTORS LATE HENRY EGLINTON.
                                   Flock No. 23.

1129   **No. 85.**
                                   Sire, No. 472
       Dam, a Whare Kauhau ewe

1130   **No. 88.**
                                   Sire, No. 472
       Dam, a Whare Kauhau ewe

1131   **No. 90.**
                                   Sire, No. 472
       Dam, a Whare Kauhau ewe

1132   **No. 93.**
                                   Sire, No. 472
       Dam, a Whare Kauhau ewe

1133   **No. 99.**
                                   Sire, No. 472
       Dam, a Whare Kauhau ewe

Flock No. 24.

Mr. GEORGE WHEELER.

Flock Book No.

1134 **ROYALIST,**
born 1904.
Bred by Mr. C. File.
(No. 15008), Vol. XI,
Kentish Flock Book.

s. Windsor 71st.
(11874)

s. Windsor 21st.
(9422)

s. Windsor 4th.
(8055)
{ s. Windsor 1st,
  6899
  d. Elham Ewe

d. Elham ewe
{ s. Jumbo, Junr.
  610
  d. Elham ewe

d. Elham ewe
No. 116

s. Jumbo 16th.
4824
{ s. Jumbo, Junr.
  610
  d. Elham ewe
  s. Boro 609

d. Elham ewe

d. Elham ewe
No. 31 6

s. Windsor 1st.
6899

s. Jumbo 15th.
4728
{ s. Jumbo, Junr.
  610
  d. Elham ewe

d. Elham ewe
{ d. Elham ewe

Mr. GEORGE WHEELER.   Flock No 24.

Flock
Book No.

1135   **Waterloo 2nd., born 1905.**

                         Sire, Waterloo (413)
Dam, Burford No. 171   ,,   Yeoman (8048)
2 d.   Burford No. 127, (34) ,,   The Shah (56)
3 d.   Propriety (14) imp.

1136   **Godolphin 3rd., born 1905**

                         Sire, Godolphin (419) imp.
Dam, Burford No. 195   ,,   Windsor 11th (8062) imp.
2 d.   Burford No. 80   ,,   Moslem (89)
3 d.   Burford No. 73   ,,   Waiorongomai (66)
4 d.   Burford No. 42   ,,   Nizam (21)
5 d.   Burford No. 18   ,,   Rector

1137   **Godolphin, 4th., born 1905.**

                         Sire, Godolphin (419) imp.
Dam, Burford No. 80   ,,   Moslem (89)
2 d.   Burford No. 73   ,,   Waiorongomai (66)
3 d.   Burford No. 42   ,,   Nizam (21)
4 d.   Burford No. 18   ,,   Rector

1138   **Watchman 4th, born 1905.**

                         Sire, Watchman (414)
Dam, a Burford ewe

1139   **Watchman 5th, born 1905.**

                         Sire, Watchman (414)
Dam, a Burford ewe

1140   **Watchman 6th, born 1905.**

                         Sire, Watchman (414)
Dam, a Burford ewe

1141   **Waterloo 3rd, born 1905.**

                         Sire, Waterloo (413)
Dam, a Burford ewe

1142   **Waterloo 4th,** born 1905.

Sire, Waterloo (413)

Dam, a Burford ewe

1143   **Waterloo 5th,** born 1905.

Sire, Waterloo (413)

Dam, a Burford ewe

1144   **Monarch 5th,** born 1905.

Sire, Monarch 3rd (422)

Dam, a Burford ewe

1145   **Monarch 6th,** born 1905.

Sire, Monarch 3rd (422)

Dam, a Burford ewe

1146   **Osborn,** born 1905.

Sire, Wimbledon (563)

Dam, a Burford ewe

1147   **Osborn 2nd,** born 1905.

Sire, Wantage (561)

Dam, a Burford ewe

1148   **Osborn 3rd,** born 1905.

Sire, Westoe (562)

Dam, a Burford ewe

1149   **Osborn 4th,** born 1905.

Sire, Westoe (562)

Dam, a Burford ewe

1150   **Ceylon,** born 1905.

Sire, Herald 2nd (552)

Dam, a Burford ewe

1151   **Burford No. 227,** born 1903.

                                  Sire, Windsor 11th (8062)

Dam, a Burford ewe

1152   **Herald 3rd,** born 1905.

                                  Sire, Herald 2nd (552)

Dam, a Burford ewe

1153   **Herald 4th,** born 1905.

                                  Sire, Herald 2nd (552)

Dam, a Burford ewe

1154   **Godfrey 2nd,** born 1905.

                                  Sire, Godolphin (419)

Dam, a Burford ewe

1155   **Godfrey 3rd,** born 1905.

                                  Sire, Godolphin (419)

Dam, a Burford ewe

1156   **Godfrey 4th,** born 1905.

                                  Sire, Godolphin (419)

Dam, a Burford ewe

1157   **Godfrey 5th,** born 1905.

                                  Sire, Godolphin (419)

Dam, a Burford ewe

1158   **Harold 3rd,** born 1905.

                                  Sire, Harold 2nd (571)

Dam, a Burford ewe

1159   **Harold 4th,** born 1905.

                                  Sire, Harold 2nd (571)

Dam, a Burford ewe

1160   **Harold 5th,** born 1905.

                           Sire, Harold 2nd (571)

Dam, a Burford ewe

1161   **Harold 6th,** born 1905.

                           Sire, Harold 2nd (571)

Dam, a Burford ewe

1162   **Rangitane,** born 1905.

         Sold to Mr. G. A. Wheeler, Kawhatau.

                    Sire, either Wimbledon (563)
                                or Wantage (561)

Dam, a Burford ewe

1163   **Burford No. 227,** born 1903.

                           Sire, Windsor 11th

Dam, a Burford ewe

Mr. WALTER C. DORSET. Flock No. 25.

1164   **E.W.D. 022,** born 1904.

              Bred by Mr. E. W. Dorset.

                      Sire, a Kahikatea ram

Dam, a Kahikatea ewe

1165   **Fairburn,** born 1904.

             Bred by Messrs. Gray Bros.

                      Sire, Duke of Kent 4th

Dam, a Fairburn ewe

Mr. E. J. RIDDIFORD.   Flock No. 27.

Flock
Book No.

1166   **Lord Roberts 11,** born 1905.

                                        Sire, Lord Roberts
          Dam, an Orongorongo ewe

1167   **Lord Roberts 28,** born 1905.

                                        Sire, Lord Roberts
          Dam, an Orongorongo ewe

1168   **Lord Roberts 33,** born 1905.

                                        Sire, Lord Roberts
          Dam, an Orongorongo ewe

1169   **Lord Roberts 8,** born 1905.

                                        Sire, Lord Roberts
          Dam, an Orongorongo ewe

1170   **Lord Roberts 20,** born 1905.

                                        Sire, Lord Roberts
          Dam, an Orongorongo ewe

Mr. T. P. ALLEN,   Flock No. 35.

1171   **Handsome Jack,** born 1905.
                    Sold to Mr. E. W. Dorset.

                                        Sire, Haldon
          Dam, a Walmer Farm ewe

1172   **Walmer I.,** born 1905.

                                        Sire, Haldon
          Dam, a Walmer Farm ewe

Mr. W. B. ALLEN.   Flock No. 36.

Flock
Book No.

1173   **Waiorongomai No. 120,** born 1904.
             Bred by Mr. A. Matthews, Waiorongomai.

                                        Sire, Windsor 83
       Dam, a Waiorongomai ewe

1174   **Waiorongomai No. 334,** born 1904.
             Bred by Mr. A. Matthews, Waiorongomai.

                                        Sire, No. 599
       Dam, a Waiorongomai ewe

             Haldon (586), born 1903.
             Bred by Messrs. Bealey Bros.

                                        Sire, Hazel
       Dam, a Haldon ewe

Messrs. GRAY BROS.   Flock No. 38.

1175   **Hororata I.,** born 1904.
             Bred by Messrs. Bealey Bros.

                                        Sire, Delphic Lad
       Dam,   Hororata ewe

1176   **Hororata II.,** born 1904.
             Bred by Messrs. Bealey Bros.

                                        Sire, Ashford Hero
       Dam,   Hororata ewe

1177   **E.W.D. 165,** born 1904.
             Bred by Mr. W. Dorset.

                                        Sire, a Kahikatea ram
       Dam, a Kahikatea ewe

Mr. DAVID BUICK.   Flock No. 42.

Flock
Book No.

1178   **Cloverlea-Walmer No. 1, born 1905.**
                     Sire, Walmer (imp.)
Dam, a Cloverlea ewe

1179   **Cloverlea-Walmer No. 2, born 1905.**
                     Sire, Walmer (imp.)
Dam, a Cloverlea ewe

1180   **Cloverlea-Walmer No. 3, born 1905.**
                     Sire, Walmer (imp.)
Dam, a Cloverlea ewe

1181   **Cloverlea-Walmer No. 4, born 1905.**
                     Sire, Walmer (imp.)
Dam, a Cloverlea ewe

1182   **Cloverlea-Walmer No. 5, born 1905.**
                     Sire, Walmer (imp.)
Dam, a Cloverlea ewe

Mr. WILLIAM F. JACOB.   Flock No. 43.

1183   **Young Godwin 1st, born 1905.**
                     Sire, Young Godwin (593)
Dam, a Kiwitea ewe

1184   **Young Godwin 2nd, born 1905.**
                     Sire, Young Godwin (593)
Dam, a Kiwitea ewe

1185   **Young Godwin 3rd, born 1905.**
                     Sire, Young Godwin (593)
Dam, a Kiwitea ewe

Flock
Book No.

1186 **Favorite,** born 1905.

Sire, Young Godwin (593)

Dam, a Kiwitea ewe

1187 **Kiwitea 14th,** born 1905.

Sire, Westbroke 1 (302)

Dam, a Kiwitea ewe

1188 **Kiwitea 15th,** born 1905.

Sire, Westbroke 1 (302)

Dam, a Kiwitea ewe

1189 **Togo 402,** born 1905.
Bred by Mr. J. O. Batchelar.

Sire, Togo (601)

Dam, a Willowbank ewe

1190 **Togo 404,** born 1905.
Bred by Mr. J. O. Batchelar.

Sire, Togo (601)

Dam, a Willowbank ewe

1191 **Togo 405,** born 1905.
Bred by Mr. J. O. Batchelar.

Sire, Togo (601)

Dam, a Willowbank ewe

1192 **Togo 408,** born 1905.
Bred by Mr. J. O. Batchelar.

Sire, Togo (601)

Dam, a Willowbank ewe

1193 **Tiritea Duke 428,** born 1905.
Bred by Mr. J. O. Batchelar.

Sire, Tiritea Duke (603)

Dam, a Willowbank ewe

Flock
Book No.

1194   **Tiritea Duke 456,** born 1905.

Bred by Mr. J. O. Batchelar.

Sire, Tiritea Duke (603)

Dam, a Willowbank ewe

Mr. G. R. SYKES.   Flock No. 48.

1195   **Hazelwood 2nd, U. 4,** born 1905.

Bred by Mr. Ernest Short.

Sire, Young Hazelwood (594)

Dam, a Parorangi ewe

Mr. H. S. HADFIELD.   Flock No. 49.

1196   **Bedrock,** born 1905.

Bred by Mr. W. B. Allen.

Sire, No. 99 (589)

Dam, a Clareville ewe

Mr. DAVID ROWLAND.   Flock No. 50.

———

Flock
Book No.

1197   **Bloomfield 12th,** born 1904.

                                Sire, Bloomfield 1st
        Dam, a Bloomfield ewe

1198   **Bloomfield 13th,** born 1904.

                                Sire, Bloomfield 1st
        Dam, a Bloomfield ewe

1199   **Bloomfield 14th,** born 1904.

                                Sire, Bloomfield 1st
        Dam, a Bloomfield ewe

1200   **Bloomfield 15th,** born 1904.

                                Sire, Bloomfield 1st
        Dam, a Bloomfield ewe

1201   **Bloomfield 16th,** born 1904.

                                Sire, Windsor 3rd
        Dam, a Bloomfield ewe

1202   **Manuka,** born 1904.
                Bred by Mr. J. O. Batchelar.

                                Sire, Prince Imperial
        Dam, a Tiritea ewe

———

Exs. late Mr. JAS. WILKINSON'S ESTATE.

                                Flock No. 57.

———

1203   **Tahora No. 33,** born 1904.
                Bred by Mr. E. Eglinton.

                                Sire, a stud ram
        Dam, a stud ewe

1204   **Tahora No. 34,** born 1904.
                       Bred by Mr. E. Eglinton.

                                          Sire, a stud ram
          Dam, a stud ewe

1205   **Tahora No. 35,** born 1904.
                       Bred by Mr. E. Eglinton.

                                          Sire, a stud ram
          Dam, a stud ewe

1206   **Tahora No. 36,** born 1904.
                       Bred by Mr. E. Eglinton.

                                          Sire, a stud ram
          Dam, a stud ewe

Mr. ROBERT TANNER.   Flock No. 63.

1207   **Bloomfield 4th** (464), born 1902.
                       Bred by Mr. D. Rowland.

                                       Sire, Windsor 3rd, imp.
          Dam, a Ripon ewe

Mr. P. MUNGAVIN.   Flock No. 64.

1208   **Sampson,** born 1904.
                   Bred by Mr. R. Tanner, Longburn.

                                       Sire, Sampson 2nd
          Dam, a Longburn ewe

1209   **Tattoo No. 5,** born 1904.

Bred by Mr. J. Holms, Waimahaka.

                           Sire, a stud ram

Dam, a Waimahaka ewe

1210   **Tattoo No. 8,** born 1904.

Bred by Mr. J. Holms, Waimahaka.

                           Sire, a stud ram

Dam, a Waimahaka ewe

1211   **Porirua 15th,** born 1904.

                           Sire, Porirua No. 13

Dam, a Willowbank ewe

Mr. HERBERT S. HAWKINS. Flock No. 65.

1212   **Glencoe No. 10,** born 1903.

                           Sire, No. 482

Dam, a Glencoe flock ewe

1213   **Glencoe No. 11,** born 1904.

                           Sire, No. 481

Dam, a Glencoe flock ewe

1214   **Glencoe No. 12,** born 1904.

                           Sire, No. 294

Dam, a Glencoe flock ewe

1215   **Glencoe No. 13,** born 1904.

                           Sire, No. 151

Dam, a Glencoe flock ewe

1216   **Glencoe No. 14,** born 1904.

                           Sire, No. 293

Dam, a Glencoe flock ewe

Messrs. F. HUTCHINSON & SON.

Flock No. 71.

————

Flock
Book No.

1217   **Cardinal VI., born 1902.**

                                         Sire, No. 109

Dam, a Rissington ewe

1218   **No. 6 of 1903.**

                                         Sire, No. 109

Dam, a Rissington ewe

1219   **No. 7 of 1903.**

                                         Sire, No. 109

Dam, a Rissington ewe

1220   **No. 27 of 1903.**

                                         . Sire, No. 109

Dam, a Rissington ewe

1221   **No. 4, born 1904.**

                                         Sire, No. 109

Dam, a Rissington ewe

1222   **Sir Arthur, born 1903.**

                                         Sire, Westbroke

Dam, a Rissington ewe

1223   **Squire Finn, born 1903.**

                                         Sire, Westbroke

Dam, a Rissington ewe

1224   **Tonbridge, born 1903.**

                                         Sire, Earl Godwin

Dam, a Rissington ewe

Mr. J. W. HARDING.   Flock No. 72.

Flock
Book No.

1225   **Mount Bealey A1,** born 1905.
                         Sire, Bealey's Pride (699)
     Dam, a Mount Vernon ewe

1226   **Mount Bealey A2,** born 1905.
                         Sire, Bealey's Pride (699)
     Dam, a Mount Vernon ewe

1227   **Mount Bealey A3,** born 1905.
                         Sire, Bealey's Pride (699)
     Dam, a Mount Vernon ewe

1228   **Mount Bealey A4,** born 1905.
                         Sire, Bealey's Pride (699)
     Dam, a Mount Vernon ewe

1229   **Mount Bealey A5,** born 1905.
                         Sire, Bealey's Pride (699)
     Dam, a Mount Vernon ewe

1230   **Mount Bealey A6,** born 1905.
                         Sire, Bealey's Pride (699)
     Dam, a Mount Vernon ewe

1231   **Mount Bealey A7,** born 1905.
                         Sire, Bealey's Pride (699)
     Dam, a Mount Vernon ewe

1232   **Mount Bealey A8,** born 1905.
                         Sire, Bealey's Pride (699)
     Dam, a Mount Vernon ewe

1233   **Mount Bealey A9,** born 1905.
                         Sire, Bealey's Pride (699)
     Dam, a Mount Vernon ewe

1234   **Mount Bealey A10,** born 1905.
                                        Sire, Bealey's Pride (699)
        Dam, a Mount Vernon ewe

1235   **Mount Bealey A11,** born 1905.
                                        Sire, Bealey's Pride (699)
        Dam, a Mount Vernon ewe

1236   **Mount Bealey A12,** born 1905.
                                        Sire, Bealey's Pride (699)
        Dam, a Mount Vernon ewe

1237   **Mount Bealey A13,** born 1905.
                                        Sire, Bealey's Pride (699)
        Dam, a Mount Vernon ewe

1238   **Mount Bealey A14,** born 1905.
                                        Sire, Bealey's Pride (699)
        Dam, a Mount Vernon ewe

1239   **Mount Bealey A15,** born 1905.
                                        Sire, Bealey's Pride (699)
        Dam, a Mount Vernon ewe

1240   **Mount Bealey A16,** born 1905.
                                        Sire, Bealey's Pride (699)
        Dam, a Mount Vernon ewe

1241   **Mount Bealey A17,** born 1905.
                                        Sire, Bealey's Pride (699)
        Dam, a Mount Vernon ewe

1242   **Mount Bealey A18,** born 1905.
                                        Sire, Bealey's Pride (699)
        Dam, a Mount Vernon ewe

1243   **Mount Vernon B2,** born 1905.

Sire, Mount Vernon 13th (628)

Dam, a Mount Vernon ewe

Mr. ERNEST SHORT.   Flock No. 77.

1244   **Reliance F66,** born 1905.

Sire, Favorite (330)

Dam, a Parorangi ewe

1245   **Record 2nd,** born 1905.

Sire, Record (663)

Dam, a Parorangi ewe

1246   **Record 3rd,** born 1905.

Sire, Record (663)

Dam, a Parorangi ewe

1247   **Record 4th,** born 1905.

Sire, Record (663)

Dam, a Parorangi ewe

1248   **Record 5th,** born 1905.

Sire, Record (663)

Dam, a Parorangi ewe

1249   **Record 6th,** born 1905.

Sire, Record (663)

Dam, a Parorangi ewe

1250   **Record 7th,** born 1905.

Sire, Record (663)

Dam, a Parorangi ewe

1251   **Record 8th,** born 1905.

                                        Sire, Record (663)

    Dam, a Parorangi ewe

1252   **Record 9th,** born 1905.
                    Sold to Mr. F. Kensington, Rewa.

                                        Sire, Record (663)

    Dam, a Parorangi ewe

1253   **Record 10th,** born 1905.
                    Sold to Mr. J. C. Field, Gisborne.

                                        Sire, Record (663)

    Dam, a Parorangi ewe

1254   **Record 11th,** born 1905.

                                        Sire, Record (663)

    Dam, a Parorangi ewe

1255   **Record 15th,** born 1905.

                                        Sire, Record (663)

    Dam, a Parorangi ewe

1256   **Superior I.12,** born 1905.

                                        Sire, Young Model (629)

    Dam, a Parorangi ewe

1257   **Goldreef I.13,** born 1905.
                    Sold to Mr. Hebden, N.S. Wales.

                                        Sire, Young Model (629)

    Dam, a Parorangi ewe

1258   **Hero I.14,** born 1905.

                                        Sire, Young Model (629)

    Dam, a Parorangi ewe

1259   **The Briton I.34,** born 1905.
                    Sold to Mr. Hebden, N.S. Wales.

                                        Sire, Young Model (629)

    Dam, a Parorangi ewe

Flock
Book No.

1260 **Combine J.20,** born, 1905.
Sold to Mr. C. A. J. Levett, Kiwitea.

Windsor 60th (501)

Dam, a Parorangi ewe

1261 **Restorer J.22,** born 1905.
Sire, Windsor 60th (501)

Dam, a Parorangi ewe

1262 **Windsor J.31,** born 1905.
Sold to Mr. T. Hunt, Nelson.

Sire, Windsor 60th (501)

Dam, a Parorangi ewe

1263 **Austral, Q.50,** born 1905.
Sold to Mr. Hebden, N.S. Wales.

Sire, Windsor 61st (500)

Dam, a Parorangi ewe

1264 **Invader R.3,** born 1905.
Sold to Mr. Hebden, N.S. Wales.

Sire, Buffalo Bill 5th (507)

Dam, a Parorangi ewe

1265 **Umpire S.4,** born 1905.
Sire, Earl Godwin (301)

Dam, a Parorangi ewe

1266 **Defiance S.12,** born 1905.
Sire, Earl Godwin (301)

Dam, a Parorangi ewe

1267 **Blondin S.13,** born 1905.
Sold to Mr. W. E. Bidwill, Featherston.

Sire, Earl Godwin (301)

Dam, a Parorangi ewe

1268   **Kentish Lad,** born 1905.
                    Sold to Mr. T. Hunt, Nelson.

                              Sire, Young Hazlewood (594)
        Dam, a Parorangi ewe

1269   **Hazlewood 2nd,** born 1905.
                    Sold to Messrs. Sykes Bros.; Masterton.

                              Sire, Young Hazlewood (594)
        Dam, a Parorangi ewe

1270   **Desirable,** born 1905.
                    Sold to Mr. Hebden, N.S. Wales.

                              Sire, Young Hazlewood (594)
        Dam, a Parorangi ewe

1271   **Treasure,** born 1905.
                    Bred by Mr. J. O. Batchelar, Palmerston North.

                              Sire, Togo (601)
        Dam, a Willowbank ewe

1272   **Sandow,** born 1904.

                              Sire, Major (323)
        Dam, a Parorangi ewe

1273   **Matchless,** born 1904.

                              Sire, Romeo (332)
        Dam, a Parorangi ewe

1274   **Comrade,** born 1901.
                    Sold to Mr. J. Campion, Waituna West.

                              Sire, Parorangi (326)
        Dam, a Parorangi ewe

1275   **Windsor 2nd,** born 1904.
                    Sold to Mr. Rui Batley, Utiku.

                              Sire, Windsor 60th (501)
        Dam, a Parorangi ewe

1276   **Ngatarua,** born 1905.
Sold to Mr. Rui Batley, Utiku.

Sire, Young Hazlewood (594)

Dam, a Parorangi ewe

1277   **Attractive,** born 1905.
Sold to Mr. John Miller, Dunedin.

Sire, Active (936)

Dam, a Parorangi ewe

1278   **Review,** born 1905.
Sold to Mr. E. Eglinton, Featherston.

Sire, Young Model (629)

Dam, a Parorangi ewe

1279   **Young Nugget 1.6,** born 1905.
Sold to Mr. J. Rigger, Hastings.

Sire, Nugget (630)

Dam, a Parorangi ewe

1280   **Referee,** born 1904.
Bred by Mr. W. Baker.

Sire, a Woodendean ram

Dam, a Woodendean ewe

Sold to Mr. W. E. Bidwill, Featherston.

1281   **Elevation,** born 1905.
Bred by Mr. E. J. Riddiford.

Sire, Lord Roberts (489)

Dam, an Orongorongo ewe

1282   **Challenger R.2,** born 1905.
Sold to Mr. R. Tanner, Longburn.

Sire, Buffalo Bill 5th (507)

Dam, a Parorangi ewe

1283   **Nutshell S.22**, born 1905.
                Sold to Mr. Alfred Matthews, Featherston.

                                    Sire, Earl Godwin, (301)
        Dam, a Parorangi ewe

1284   **Renown Q.8**, born 1905.
                                    Sire, Windsor 61st (500)
        Dam, a Parorangi ewe

1285   **Conservation T.15**, born 1905.
                                    Sire, Active (936)
        Dam, a Parorangi ewe

1286   **Elegant F.8**, born 1905.
                                    Sire, Favorite (330)
        Dam, a Parorangi ewe

1287   **Dictator R.14**, born 1905.
                                    Sire, Buffalo Bill 5th (507)
        Dam, a Parorangi ewe

1288   **Robust I.1**, born 1905.
                                    Sire, Nugget (630)
        Dam, a Parorangi ewe

1289   **Invincible F.1**, born 1905.
                                    Sire, Favorite (330)
        Dam, a Parorangi ewe

1290   **Inspector F.15**, born 1905.
                                    Sire, Favorite (330)
        Dam, a Parorangi ewe

1291   **Reformer F.60**, born 1905.
                                    Sire, Major (323)
        Dam, a Parorangi ewe

Flock
Book No.

1292   **Fascinator S.14,** born 1905.

                                        Sire, Earl Godwin (301)

Dam, a Parorangi ewe

1293   **Excelsior R.13,** born 1905.

                                        Sire, Buffalo Bill 5th (507)

Dam, a Parorangi ewe

1294   **Liberator I.35,** born 1905.

                                        Sire, Nugget (630)

Dam, a Parorangi ewe

1295   **Darwin S.17,** born 1905.
                     Sold to Mr. J. Guylee.

                                        Sire, Earl Godwin (301)

Dam, a Parorangi ewe

1296   **Active 2nd T.7,** born 1905.
                     Sold to Mr. F. Kensington, Rewa.
                                        Sire, Active (936)

Dam, a Parorangi ewe

1297   **Active 3rd T.18,** born 1905.
                     Sold to Mr. D. G. Forlong.

                                        Sire, Active (936)

Dam, a Parorangi ewe

Mr. H. N. WATSON.   Flock No. 82.

1298   **Westbroke No. 17 of 1904.**
           Bred by Mr. Arthur Finn, Westbroke, Lydd, Kent.

                                        Sire, Westbroke 85 of 1901.

Dam, a Westbroke ewe

Mr. GEORGE W. BULL.   Flock No. 83.

Flock
Book No.

Young Godwin 1st (1183), born 1905.
Bred by Mr. W. F. Jacob, Kiwitea.

Sire, Young Godwin (593)

Dam, a Kiwitea ewe

Mr. EDWARD WILSON.   Flock No. 84.

1300   **H (in a diamond) 50/05, born 1905.**
Bred by Mr. H. S. Hadfield.

Sire, Lost Chord (565)

Dam, a Lindale ewe

Mr. FRANK KENSINGTON.   Flock No. 85.

1301   **Jimmie.**

Sire, Rakatuma

Dam, Rewarewa ewe

Mr. J. C. ALLEN.   Flock No. 86.

1302   **Annandale III., born 1905.**

Sire, Gordon I.

Dam, a Woodlands ewe

Flock
Book No.

1303   **Piako I.,** born 1906.

Sire, Gordon I.

Dam, an Annandale ewe

1304   **Piako II.,** born 1906.

Sire, Gordon I.

Dam, an Annandale ewe

Mr. DAVID PENRUDDOCKE BUCHANAN.

Flock No. 91.

1305   **Mayfield 18th,** born 1905.

Sire, Windermere (415)

Dam, No. 135 (42)

1306   **Mayfield 19th,** born 1905.

Sire, Windermere (415)

Dam No. 154 (60)

1307   **Mayfield 20th,** born 1905.

Sire, Windermere (415)

Dam, No. 166 (72)

1308   **Mayfield 21st,** born 1905.

Sire, Windermere (415)

Dam, No. 166 (72)

1309   **Mayfield 23rd,** born 1905.

Sire, Windermere (415)

Dam, No. 156 (62)

1310   **Mayfield 24th,** born 1905.

Sire, Windermere (415)

Dam, No. 156 (62)

1311   **Mayfield 25th,** born 1905.

                                  Sire, Windermere (415)

      Dam, No. 137 (44)

Mr. A. R. FANNIN.   Flock No. 93.

1312   **Dealwood 4th,** born 1904.
                      Bred by Mr. W. B. Allen.

                            Sire, bred by Mr. J. Holms

      Dam, a Dealwood ewe

Mr. J. C. FIELD.   Flock No. 94.

1313   **No. 378,** born 1900.
                      Bred by Mr. J. F. Reid.

                                  Sire, a stud ram

      Dam, an Elderslie ewe

Mr. J. E. HEWITT.   Flock No. 96.

1314   **Porirua 6th,** born 1905.
                      Bred by Mr. P. Mungavin.

                                  Sire, Porirua 13th

      Dam, a Willowbank ewe

Mr. JAMES KNIGHT.   Flock No. 97.

Flock
Book No.

The Earl (1318), born 1905.
Bred by Mr. E. Short.

Dam, a Woodendean ewe

Sire, The Duke (935)

Messrs. McKENZIE & LOVELOCK.

Flock No. 101.

1315

Bealey No. 1, born 1905.
Bred by Messrs. Bealey Bros.

Sire, Hazle XV. (1521, Vol 2) imp.

Dam, a Bealey ewe

1316

Bealey No. 2, born 1904.
Bred by Messrs. Bealey Bros.

Sire, Rigden's No. 112 of 1901, (10637 Vol. 8), imp.

Dam, a Bealey ewe

Mr. W. G. PEARCE.   Flock No. 104.

1317

The Nut, born 1904.
Bred by Mr. Robert Tanner, Longburn.

Sire, a ram bred by Mr. J. F. Reid

Dam, Longburn ewe   ,,   Yanko by Cardinal (imp.)

Mr. ERNEST SHORT.   Flock No. 109.

Flock
Book No.

1318   **The Earl H.1,** born 1905.
                 Sold to Mr. James Knight, Feilding.

                                    Sire, The Duke (935)
      Dam, a Woodendean ewe

1319   **Viscount H.3,** born 1905.
                                    Sire, The Duke (935)
      Dam, a Woodendean ewe

1320   **Marquis H.5,** born 1905.
               Sold to Mr. Hebden, New South Wales.

                                    Sire, The Duke (935)
      Dam, a Woodendean ewe

1321   **Handsome H.9,** born 1905.
              Sold to Mr. W. E. Bidwill, Featherston.

                                    Sire, The Duke (935)
      Dam, a Woodendean ewe

1322   **Baronet H.14,** born 1905.
                                    Sire, The Duke (935)
      Dam, a Woodendean ewe

1323   **Kenilworth H.21,** born 1905.
                 Sold to Mr. John Allen, Waingaro.

                                    Sire, The Duke (935)
      Dam, a Woodendean ewe

1324   **Officer H.29,** born 1905.
               Sold to Mr. E. Eglinton, Featherston.

                                    Sire, The Duke (935)
      Dam, a Woodendean ewe

1325   **Success H.19,** born 1905.
              Sold to Mr. W. E. Bidwill, Featherston.

                                    Sire, The Duke (935)
      Dam, a Woodendean ewe

Messrs. SMITH BROS. Flock No. 111.

Flock
Book No.

1326   **Colyton No. 1, born 1906.**

Sire, J.O.B. 338, G.

Dam, a Woodendean ewe

1327   **Colyton No. 2, born 1906.**

Sire, J.O.B. 338, G.

Dam, a Woodendean ewe

1328   **Colyton No. 3, born 1906.**

Sire, J.O.B. 338, G.

Dam, a Woodendean ewe

1329   **J.O.B. 338, G., born 1904.**
Bred by Mr. J. O. Batchelar.

Sire, a stud ram

Dam, a Willowbank ewe

Mr. T. ALFRED SMITH. Flock No. 112.

1330   **T.A.S. No. 6, born 1906.**

Sire, Willis (993)

Dam, a stud ewe

1331   **T.A.S. No. 7, born 1906.**

Sire, Willis (993)

Dam, a stud ewe

1332   **T.A.S, No. 8, born 1906.**

Sire, Willis (993)

Dam, a stud ewe

Mr. MAX VOSS.   Flock No. 115.

Flock
Book No.

1333   **Elderslie 2nd, No. 1905,** born 1904.
              Bred by Mr. Robt. Tanner.

                                        Sire, Lord Elderslie
      Dam, a Longburn ewe

Mr. JAMES G. WILSON.   Flock No. 119.

1334   **Willowbank,** born 1904.
              Bred by Mr. J. O. Batchelar.

                                        Sire, an Allen ram
      Dam, a Willowbank ewe

1335   **Old Constitution,** born 1903.
              Bred by Mr. J. O. Batchelar.

                                        Sire, a Willowbank ram
      Dam, a Willowbank ewe

Mr. W. WINDLEY.   Flock No. 120.

1336   **No. 113,** born 1905.
              Bred by Mr. H. S. Hadfield.

                                        Sire, Lost Chord
      Dam, a Lindale ewe

1337   **Burford No. 230,** born 1903.
              Bred by Mr. G. Wheeler.

                                        Sire, Windsor 11th (8062)
      Dam, a Burford ewe

Mr. ALFRED E. HARDING. Flock No. 121.

Flock
Book No.

1338   **Aoroa 2nd B40,** born 1902.

Sire, Bealey's Pride (699)

Dam, an Aoroa ewe

1339   **Porirua,** born 1902.
Bred by Mr. P. Mungavin.

Sire, a Porirua ram

Dam, a Porirua ewe

1340   **Aoroa 3rd B35,** born 1905.

Sire, Favourite 2nd (635)

Dam, an Aoroa ewe

Messrs. DUNCAN & CAMPION.

Flock No. 124.

1341   **Comrade No. 47,** born 1901.
Bred by Mr. E. Short.

Sire, Parorangi (326)

Dam, a Parorangi ewe

Mr. W. A. ELLIS.   Flock No. 125.

1342   **Loseley,** born 1904.
Bred by Mr. R. Tanner.

Sire, an Elderslie ram

Dam, a Longburn ewe

Mr. WILLIAM GIBSON.   Flock No. 126.

Flock
Book No.

1343   **Beck Dyke,** born 1904.
                        Bred by Mr. Robt. Cobb.

                                        Sire, Buffalo Bill 5th (507)
        Dam, a Cobb ewe

1344   **Milford King,** born 1905.
                        Bred by Mr. J. O. Batchelar.

                                        Sire, a Tiritea ram
        Dam, a Tiritea ewe

                Sold to Mr. D. H. Kilgour.

1345   **Milford Prince,** born 1905.
                        Bred by Mr. J. O. Batchelar.

                                        Sire, a Tiritea ram
        Dam, a Tiritea ewe

                Sold to Mr. R. O. French.

1346   **Milford Duke,** born 1905.
                        Bred by Mr. J. O. Batchelar.

                                        Sire, a Tiritea ram
        Dam, a Tiritea ewe

                Sold to Mrs. B. H. Slack.

1347   **Milford Wonder,** born 1905.
                        Bred by Mr. J. O. Batchelar.

                                        Sire, a Tiritea ram
        Dam, a Tiritea ewe

                Sold to Mr W. Fergusson.

1348   **Milford Model,** born 1905.
                        Bred by Mr. J. O. Batchelar.

                                        Sire, a Tiritea ram
        Dam, a Tiritea ewe

Mr. J. C. KELLY.   Flock No. 127.

Flock
Book No.

1349   **Tokomaru,** born 1903.
                       Bred by Mr. R. Cobb.

                                        Sire, a stud ram

         Dam, a Cobb ewe

Mr. F. W. H. KUMMER.   Flock No. 128.

1350   **Mahoe I.,** born 1905.
                       Bred by Mrs. B. H. Slack.

                                        Sire, King Cobb

         Dam, a Utiku ewe

Mr. MAURICE MASON.   Flock No. 131.

1351   **G. 307,** born 1902.
                       Bred by Mr. J. O. Batchelar.

                                   Sire, Prince Imperial (176)

         Dam, a Willowbank ewe

Mr. O. M. MONCKTON.   Flock No. 132.

1352   **Rissington,** born 1901.
                    Bred by Messrs. F. Hutchinson & Son.

                                        Sire, Cardinal 2nd

         Dam, a Rissington ewe

Flock
Book No.
1353   **Fairfield.**
                      Bred by Mr. H. N. Watson.

                                        Sire, a stud ram
    Dam, a stud ewe

Mr. J. G. OATES.   Flock No. 133.

1354   **E.W.D. 026, born 1903.**
                       Bred by Mr. E. W. Dorset.

                                   Sire, a Kahikatea ram
        Dam, a Kahikatea ewe

Mr. W. B. V. PEARCE.   Flock No. 134.

1355   **Utiku.**
                        Bred by Mr. R. Cobb.

                                     Sire, a stud ram
        Dam, a Cobb ewe

Mr. GERALD TOLHURST.   Flock No. 138.

1356   **Rakana 1, born 1905.**

                                   Sire, McHardy's 310
    Dam, a Beaulieu ewe

1357   **Rakana 2,** born 1905.

                                        Sire, McHardy's 377
Dam, a Beaulieu ewe

Mr. W. G. AITKEN.   Flock No. 140.

1358   **Fairburn of 1905.**
                        Bred by Messrs. Gray Bros.

                                        Sire, Duke of Kent
            Dam, a Fairburn ewe

Mr. Wm. E. BIDWILL.   Flock No. 146.

1359   **Parorangi No. 897,** born 1904.
                        Bred by Mr. E. Short.

                                        Sire, a stud ram
            Dam, a Parorangi ewe

Mr. W. H. BOOTH.   Flock No. 147.

1360   **No. 1 Gray,** born 1904.
                        Bred by Messrs. Gray Bros.

                                        Sire, a stud ram
Dam, a Fairburn ewe

Mr. EWEN A. CAMPBELL.   Flock No. 149.

1361   **Brandon Hall, No. 87.**
                        Bred by Mr. James Bell.

                                          Sire, a stud ram
         Dam, a Brandon Hall ewe

Mr. E. EAGLE, Junr.   Flock No. 153.

1362   **Belvedere, born 1902.**
                        Bred by Mr. A. Matthews.

                                          Sire, a stud ram
         Dam, a Waiorongomai ewe

Mr. W. J. A. McGREGOR.   Flock No. 161.

1363   **Westwood No. 27 of 1905.**
                   Bred by Messrs. L. H. and G. W. Finn, Kent.

                                     Sire, Westbroke (13547) Vol. 10
         Dam, a Westwood ewe

EWES.

Mr. GEORGE WHEELER. Flock No. 24.

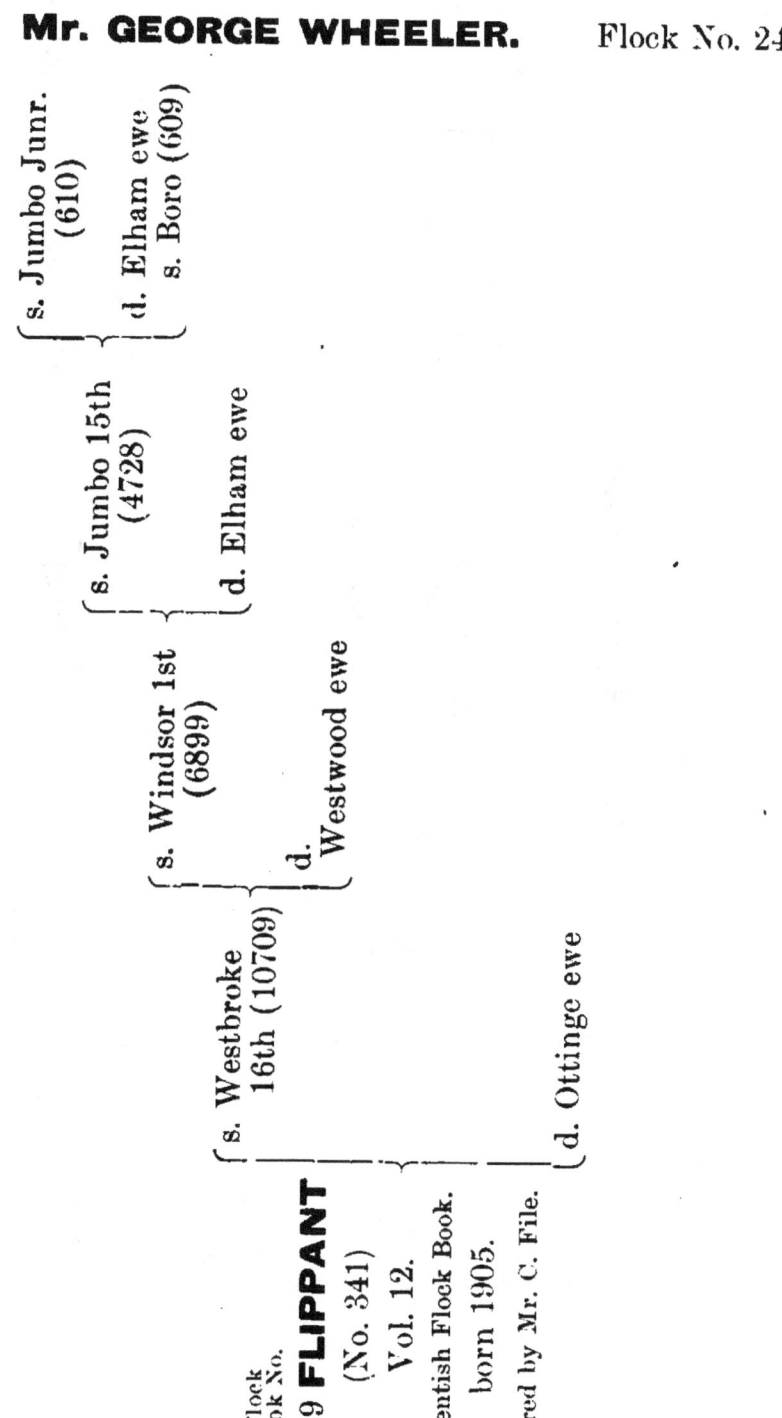

Flock Book No.
159 FLIPPANT
(No. 341)
Vol. 12.
Kentish Flock Book.
born 1905.
Bred by Mr. C. File.

s. Westbroke 16th (10709)

d. Ottinge ewe

s. Windsor 1st (6899)

d. Westwood ewe

s. Jumbo 15th (4728)

d. Elham ewe

s. Jumbo Junr. (610)

d. Elham ewe
s. Boro (509)

Mr. GEORGE WHEELER.

Flock No. 24.

Flock Book No.
160 **FILIGREE,**
(No. 342) Vol. 12.
Kentish Flock Book.
born 1905.
Bred by Mr. C. File.

{
- s. Elham No. 40 (13186)
  {
  - s. Windsor 30th. (10641)
    {
    - s. Windsor 1st, (6899)
    - d. Elham ewe
    }
  - d. Elham ewe No. 830
    {
    - s. Jumbo, Junr. (610)
    - d. Elham Ewe
    }
  }
- d. Elham ewe No. 407
  {
  - s. Windsor 21st. (9422)
    {
    - s. Windsor 4th. (8055)
    - d. Elham ewe
    }
  - d. Elham ewe No. 29
    {
    - s. Windsor 1st. (6899)
    - d. Elham ewe
    }
  }
}

Mr. GEORGE WHEELER. Flock No. 24.

Flock
Book No.

161 **Burford No. 220,** born 1905.

Sire, Godolphin (419)
Dam, Burford No. 186 (99)   s. Windsor 11th (8062)
2 d. Burford No. 139 (46)   s. Waiorongomai (66)
3 d. Fidelity (13) imp.

162 **Burford No. 221,** born 1905.

Sire, Godolphin (419)
Dam, Burford No. 168 (74)   s. Yeoman (8048)
2 d. Burford No. 120 (27)   s. Moghul (19)
3 d. The Abbess (5) imp.

163 **Burford No. 222,** born 1905.

Sire, Godolphin (419)
Dam, Burford No. 189 (102)   s. Windsor 11th (8062)
2 d. Burford No. 135 (42)   s. Waiorongomai (66)
3 d. Burford No. 119 (26)   s. Moghul (19)
4 d. The Nun (6) imp.

164 **Burford No. 223 (twin No. 229),** born 1905.

Sire, Waterloo (413)
Dam, Burford No. 67   s. Waiorongomai (66)
2 d. Burford No. 41   s. Nizam (21)
3 d. Burford No. 16   s. Osman

165 **Burford No. 224,** born 1905.

Sire, Worcester (424B)
Dam, Burford No. 183 (88)   s. Yeoman (8048)
2 d. Burford No. 113 (20)   s. Nizam (21)
3 d. The Novice (7) imp.

166 **Burford No. 225,** born 1905.

Sire, Godolphin (419)
Dam, Burford No. 201 (115)   s. Musketeer (153)
2 d. Burford No. 179 (84)   s. Yeoman (8048)
3 d. Burford No. 133 (40)   s. Waiorongomai (66)
4 d. Exile (10) imp.

167   **Burford No. 227,** born 1905.

          Sire, Godolphin (419)
Dam, Burford No. 191 (104) s. Windsor 11th (8062)
2 d. Burford No. 67   s. Waiorongomai (66)
3 d. Burford No. 41   s. Nizam (21)
4 d. Burford No. 16   s. Osman

168   **Burford No. 228,** born 1905.

          Sire, Godolphin (419)
Dam, Burford No. 194 (107) s. Windsor 11th (8062)
2 d. Burford No. 74   s. Waiorongomai (66)
3 d. Burford No. 49   s. Homespun (35)

169   **Burford No. 229 (a twin),** born 1905.

          Sire, Waterloo (413)
Dam, Burford No. 67   s. Waiorongomai (66)
2 d. Burford No. 41   s. Nizam (21)
3 d. Burford No. 61   s. Osman

170   **Burford No. 230,** born 1905.

          Sire, Godolphin (419)
Dam, Burford No. 182 (87) s. Yeoman (8048)
2 d. Burford No. 128 (35) s. The Shah (56)
3 d. The Abbess (5) imp.

171   **Burford No. 231,** born 1905.

          Sire, Godolphin (419)
Dam, Burford No. 190 (103) s. Windsor 11th (8062)
2 d. Burford No. 146 (53) s. Waiorongomai (66)
3 d. Burford No. 124 (31) s. The Shah (56)
4 d. Burford No. 112 (19) s. Nizam (21)
5 d. The Nun (6) imp.

172   **Burford No. 232,** born 1905.

          Sire, Waterloo (413)
Dam, Burford No. 172 (78) s. Yeoman (8048)
2 d. Burford No. 125 (32) s. Moghul (19)
3 d. Fidelity (13) imp.

173   **Burford No. 233,** born 1905.

                                        Sire, Godolphin (419)
Dam, Burford No. 185 (98)   s. Windsor 11th (8062)
2 d. Burford No. 145 (52)   s. Waiorongomai (66)
3 d. Burford No. 130 (37)   s. The Shah (56)
4 d. Burford No. 116 (23)   s. Rajah (19A)
5 d. Fidelity (13) imp.

174   **Burford No. 234,** born 1905.

                                        Sire, Worcester (424B)
Dam, Burford No. 184 (97)   s. Windsor 11th (8062)
2 d. Burford No. 140 (47)   s. Waiorongomai (66)
3 d. Fidelity (13) imp.

Mr. DAVID PENRUDDOCKE BUCHANAN.

Flock No. 91.

175   **Mayfield No. 26,** born 1905.

                                        Sire, Windermere (415)
Dam, No. 138 (45)   s. Waiorongomai (66)
2 d. Peeress (9) imp.   s. Proof

176   **Mayfield No. 27,** born 1905.

                                        Sire, Windermere (415)
Dam, No. 143 (50)   s. Waiorongomai (66)
2 d. No. 120 (27)   s. The Moghul (19)
3 d. The Abbess (5) imp.   s. Proof

177   **Mayfield No. 28 (a twin),** born 1905.

                                        Sire, Windermere (145)
Dam, No. 160 (66)   s. Moslem (89)
2 d. No. 144 (51)   s. Waiorongomai (66)
3 d. No. 120 (27)   s. The Moghul (19)
4 d. The Abbess (5) imp.   s. Proof

178   **Mayfield No. 29,** born 1905.

                                        Sire, Windermere (415)
        Dam, No. 151 (58)   s. Moslem (89)
        2 d. No. 124 (31)   s. The Shah (56)
        3 d. No. 112 (19)   s. Nizam (21)
        4 d. The Nun (6) imp.   s. Proof

179   **Mayfield No. 30. (a twin),** born 1905.

                                        Sire, Windermere (415)
        Dam, No. 137 (44)   . s. Waiorongomai (66)
        2 d. Peeress (9) imp.   s. Proof

180   **Mayfield No. 31,** born 1905.

                                        Sire, Windermere (415)
        Dam, No. 144 (51)   s. Waiorongomai (66)
        2 d. No. 120 (27)   s. The Moghul (19)
        3 d. The Abbess (5) imp.   s. Proof

181   **Mayfield No. 32,** born 1905.

                                        Sire, Windermere (415)
        Dam, No. 152 (59)   s. Moslem (89)
        2 d. No. 128 (35)   s. The Shah (56)
        3 d. The Abbess (5) imp.   s. Proof

Mr. W. A. ELLIS.   Flock No. 125.

182   **Willowbank No. 271G.**
                Bred by Mr. J. O. Batchelar.

                                        Sire, Prince Imperial (176)
        Dam, No. 122 G   s. Cobham (242)

Willowbank No. 118G (125).
Bred by Mr. J. O. Batchelar.

Sire, Cobham (242)
Dam, No. 23G

183   **Willowbank No. 27D.**
Bred by Mr. J. O. Batchelar.

Sire, Young Allen (244)
Dam, a Willowbank ewe

184   **Willowbank No. 268G.**
Bred by Mr. J. O. Batchelar.

Sire, Prince Imperial (176)
Dam, No. 331

185   **Willowbank No. 311G.**
Bred by Mr. J. O. Batchelar.

Sire, Prince Imperial (176)
Dam, No. 6G

186   **Willowbank No. 287G.**
Bred by Mr. J. O. Batchelar.

Sire, Prince Imperial (176)
Dam, Lost Clip   s. Cobham (242)

187   **Willowbank No. 93A.**
Bred by Mr. J. O. Batchelar.

Sire, Saturn, bred by Mr. Reid, Elderslie
Dam, a Willowbank ewe

188   **Willowbank No. 105.**
Bred by Mr. J. O. Batchelar.

Sire, Young Sutton
Dam out of Cobb ewe   s. Young Allen (244)

189   **Willowbank No. 401.**

Bred by Mr J. O. Batchelar.

Sire, Prince Imperial (176)

Dam, No. 396   s. Young Sutton

190   **Willowbank No. 142.**

Bred by Mr. J. O. Batchelar.

Sire, No. 101

Dam, No. 62   s. No. 3D

191   **Willowbank No. 410.**

Bred by Mr. J. O. Batchelar.

Sire, No. 2C

Dam, No. 97D   s. Young Sutton

ADDENDA.

Mr. ERNEST SHORT. Flock No. 109.

Flock
Book No.

1364 **Normanby H.7,** born 1905.

Sire, The Duke (935)

Dam, a Woodendean ewe

1365 **Ranfurly H.15,** born 1905.

Sire, The Duke (935)

Dam, a Woodendean ewe

1366 **Joker H.18,** born 1905.

Sold to Messrs. Barnard, Brown & Co.

Sire, The Duke (935)

Dam, a Woodendean ewe

1367 **Consul H.24,** born 1905.

Sire, The Duke (935)

Dam, a Woodendean ewe

1368 **Redoubtable H.26,** born 1905.

Sire, The Duke (935)

Dam, a Woodendean ewe

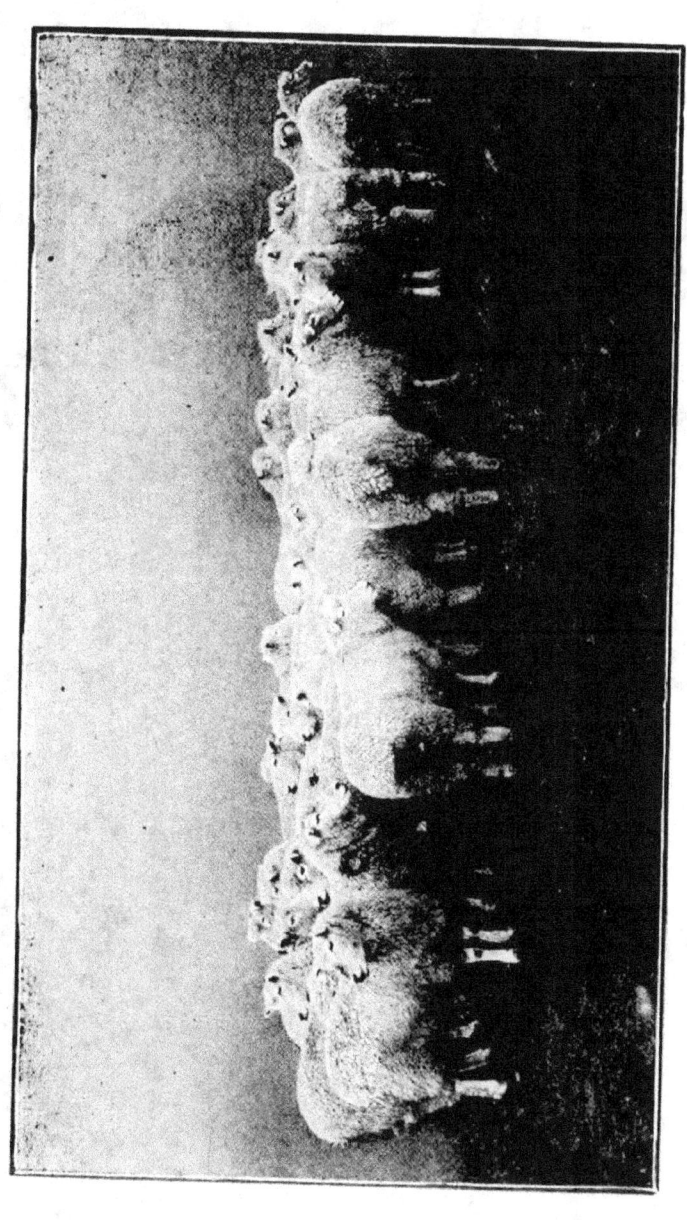

GROUP OF EWES.

THE PROPERTY OF MR. G. C. WHEELER, BURFORD, HALCOMBE.

GROUP OF STUD ROMNEY EWES.

Bred by and the Property of Mr. E. Short, Parorangi, Feilding.

"PRINCESS ROYAL."

BRED BY AND THE PROPERTY OF MR. E. SHORT, PARORANGI, FEILDING.

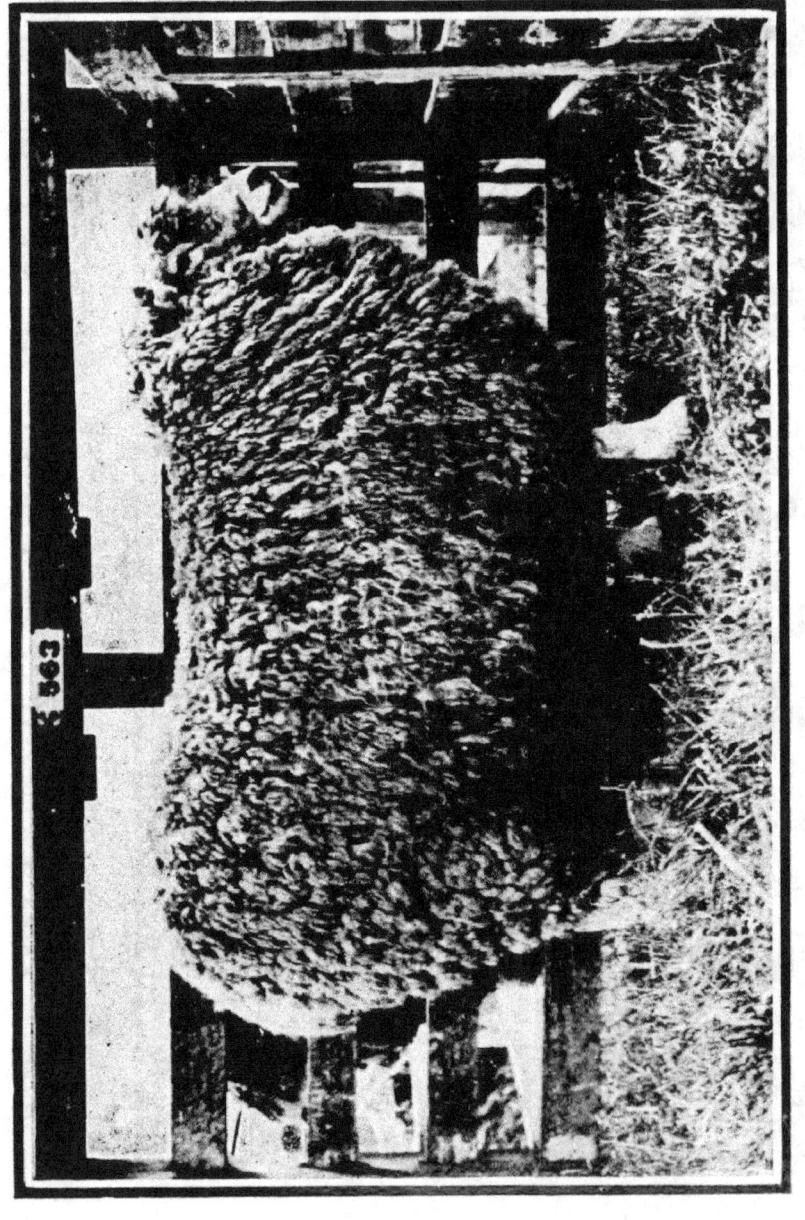

"THE MODEL 987," (943).

BRED BY AND THE PROPERTY OF MR. E. SHORT, PARORANGI, FEILDING.

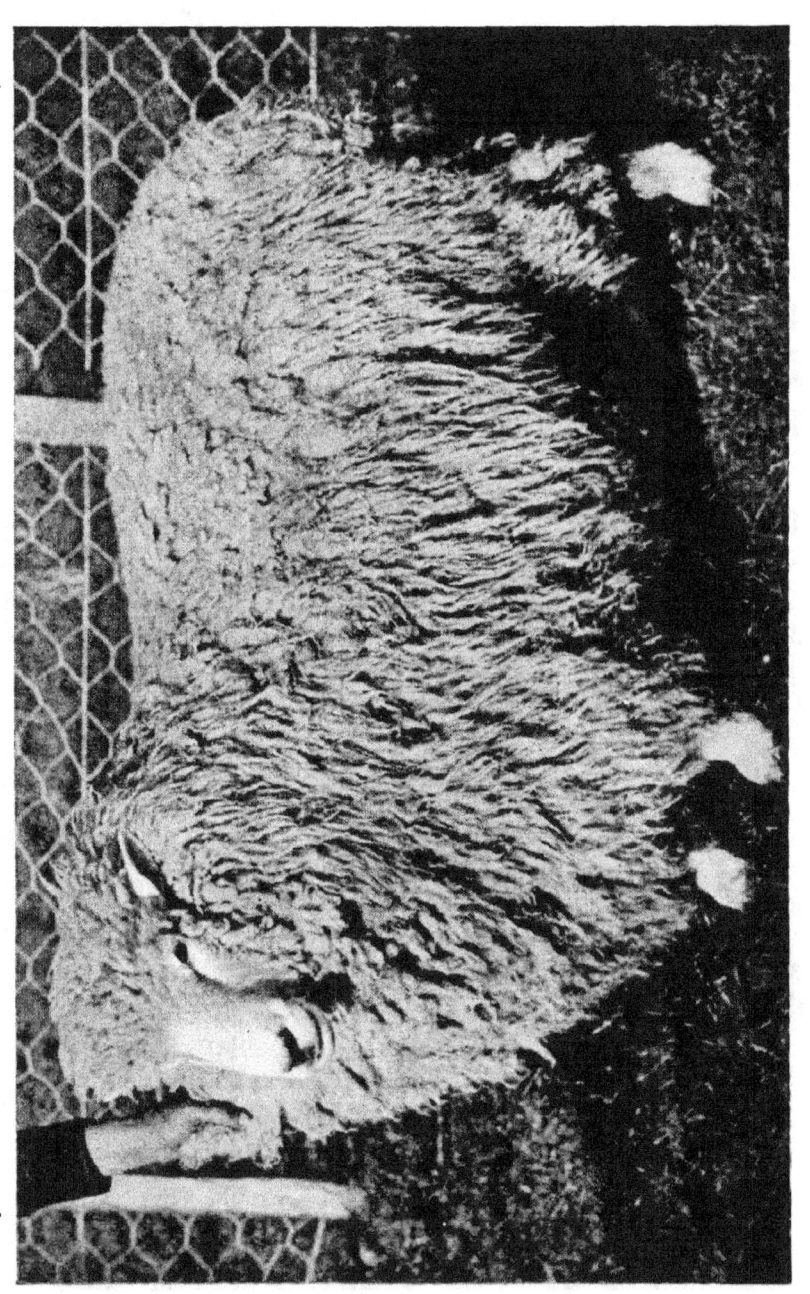

"RECORD 2ND."

BRED BY AND THE PROPERTY OF MR. E. SHORT, PARORANGI, FEILDING.

"ROYALIST." (1134)

Bred by Mr. C. File, Elham, Kent, England.

IMPORTED BY MR. G. C. WHEELER,

BURFORD, HALCOMBE.

www.ingramcontent.com/pod-product-compliance
Lightning Source LLC
Chambersburg PA
CBHW081726220526

45468CB00008B/1991